"*Go Like Hell* is a wonder, chock-a-block with great he
a pedal-to-the-metal account of greed and gumpti
Don't worry about that seat belt. Just go for the r
—Leigh Montville, AUTHOR OF *At the Altar of Spee*

Go Like Hell

Ford, Ferrari, and Their Battle for Speed and Glory at Le Mans

A. J. BAIME

PRAISE FOR

Go Like Hell

"A sophisticated tale . . . [Juxtaposes] the behind-the-scenes technology and the thrill of speed with the public tragedies they produced."

—*New York Times*

"Engaging . . . Grips you from the early pages to the conclusion."

—*AutoWeek*

"Insightful, well written accounts of the events and people involved along with inspired detail regarding the vehicles makes for a page-turner. This is an ideal book for gear-heads, automotive enthusiasts, historians, and people who might find amazing symmetry in what happened over 40 years ago versus what is happening today."

—*Denver Examiner*

"Reads like a suspense novel. Baime's exhaustive research pays off."

—*Motor Trend*

"Henry Ford II's monumental effort to topple Enzo Ferrari from the summit of sports-car racing at Le Mans is vibrantly told in this fast-paced account of the clash between the two fearsome, hyper-competitive automotive titans."

—Bloomberg

"An adrenaline-fueled jump into the pit stops of Le Mans, circa the 1960s."

—*Maxim*

"Deftly combines pacing, suspense, and beautifully constructed prose to describe real events and characters . . . Reminds us of what's possible when leaders focus resources and commitment on a goal, and refuse to fail."

—*Sports Car Digest*

"[*Go Like Hell*] belongs on any car guy's shelf next to the Haynes manual and a coffee can full of bolts and washers."

—Jalopnik

"This is truly the very best book about racing I have ever come across. Even though it's about real events, it reads like a novel."

—Art Evans, *Vintage Racecar*

"All I can say is: Wow! . . . Baime's exceptional voice puts the reader into minds of the drivers, designers, and executives who formed the Golden Age of racing; his fantastic descriptions allow the reader to feel the pounding of the cylinders as the race cars rocket down the straights and through the esses of the circuit."

—Garth Stein, author of *The Art of Racing in the Rain*

"*Go Like Hell* is an epic. Ambitions, lives, fortunes, friendship, and a place in history—all are on the line in this tale of Ford versus Ferrari. A. J. Baime marvelously carries the reader into the pits and onto the race tracks, but more important, he reveals the people behind the machines."

—Neal Bascomb, author of *The Perfect Mile* and *Hunting Eichmann*

GO LIKE HELL

Books by A.J. Baime

Big Shots: The Men Behind the Booze

Go Like Hell: Ford, Ferrari, and Their Battle for Speed and Glory at Le Mans

GO LIKE HELL

FORD, FERRARI, AND THEIR BATTLE FOR SPEED AND GLORY AT LE MANS

A. J. Baime

MARINER BOOKS
HOUGHTON MIFFLIN HARCOURT
BOSTON • NEW YORK

First Mariner Books edition 2010

www.hmhco.com

Library of Congress Cataloging-in-Publication Data

Baime, A. J. (Albert J.)
Go like hell: Ford, Ferrari, and their battle for speed and glory at Le Mans / A. J. Baime.
p. cm.
Includes bibliographical references and index.
ISBN 978-0-618-82219-5
1. Grand Prix racing—History. 2. Sports cars—United States—History. 3. Ford, Henry, 1863–1947. 4. Iacocca, Lee A. 5. Shelby, Carroll, 1923– 6. Industrialists—United States—History. 7. Automobile engineers—United States—History. 8. Automobiles—United States—Design and construction—History. 9. Sports cars—United States—Design and construction. 10. Ford Motor Company—History. 11. Ferrari automobile—History. I. Title.
GV1029.15.B35 2009
796.7'2'094417—dc22 2008052948
ISBN 978-0-547-33605-3 (pbk.)

Book design by Brian Moore

Printed in the United States of America

DOC 15 14

4500790293

I believe that if a man wanted to walk on water, and was prepared to give up everything else in life, he could do it. He could walk on water. I'm serious.

—Stirling Moss, race car driver, early 1960s

However one looks at it, Ford of Dearborn has set the cat among the pigeons. We are on the threshold of possibly the most exciting racing era in history.

—*Sports Illustrated,* May 11, 1964

CONTENTS

PART III: SPEED RISING

PART IV: THE RECKONING

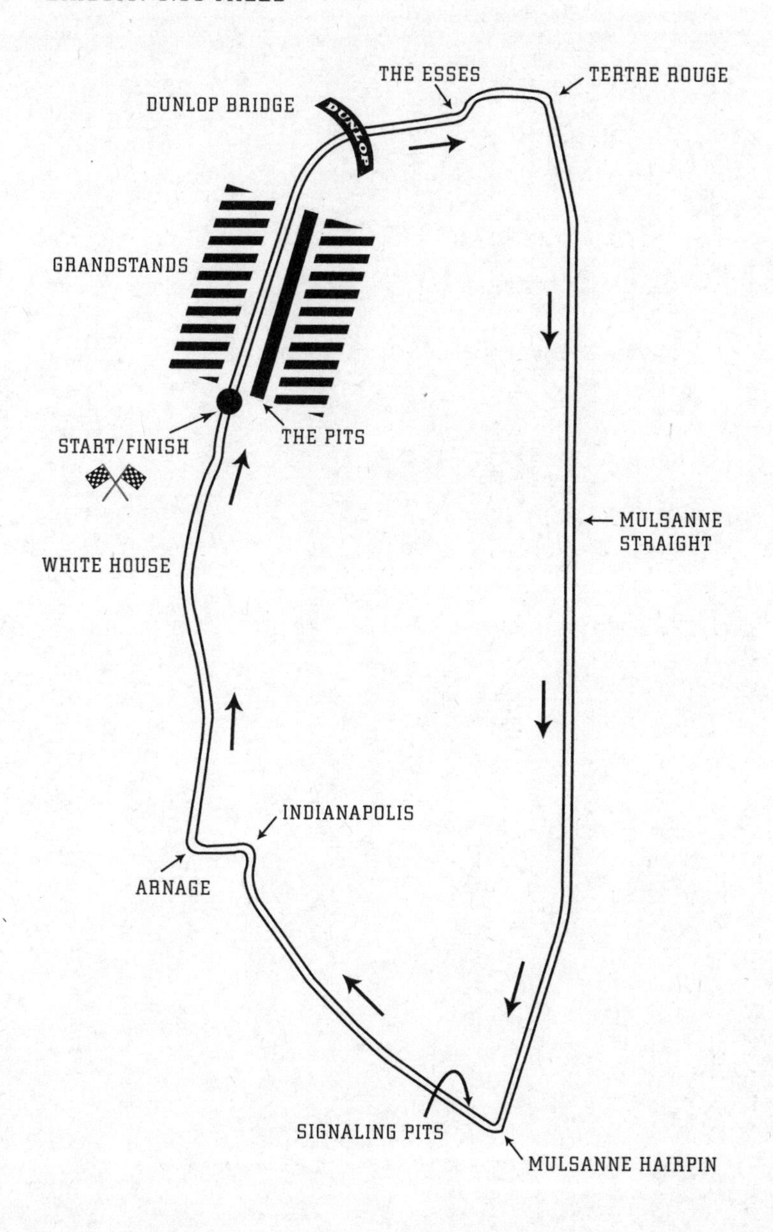
THE 24 HOURS OF LE MANS 1966
CIRCUIT: 8.36 MILES
THE ESSES
TERTRE ROUGE
DUNLOP BRIDGE
DUNLOP
GRANDSTANDS
START/FINISH
THE PITS
MULSANNE
STRAIGHT
WHITE HOUSE
INDIANAPOLIS
ARNAGE
SIGNALING PITS
MULSANNE HAIRPIN

INTRODUCTION

In 1963, following a business deal gone sour, two industrialists from either side of the Atlantic became embroiled in a rivalry that was played out at the greatest automobile race in the world. In its broad strokes, this book chronicles a clash of two titans—Henry Ford II of America and Enzo Ferrari of Italy—at the 24 Hours of Le Mans. In its finer lines, the story is about the drivers who competed and the cars they raced to victory and, in some cases, to their doom.

The men whose names will appear form a list of automotive icons: Henry Ford II, Enzo Ferrari, Lee Iacocca, Carroll Shelby, Phil Hill, John Surtees, Ken Miles, Dan Gurney, Bruce McLaren, and a rookie named Mario Andretti. Equally as important is the automobile that is born in these pages: the Ford GT, a racing car that, more than forty years after it first made its mark, is still an automobile magazine cover staple. The car was designed and built for one reason: to beat the blood-red Ferraris on their home turf, during a time when Enzo Ferrari was enjoying the greatest Le Mans dynasty in history.

The 24 Hours of Le Mans was (and still is) a sports car race. But in the 1950s and 1960s, it was more than that. It was the most magnificent marketing tool the sports car industry had ever known. Renowned manufacturers built street-legal machines that would prove on the racetrack that their cars were the best in the world. Sports car races were as beautiful as they were dangerous, and none of them was more so than Le Mans. In 1964, the first year Henry Ford II fielded a car at the 24-hour classic, *Car and Driver* magazine called the event "a four hour sprint race followed by a 20 hour death watch." It was "probably the most dangerous sporting event in the world." A win translated into millions in sales. It was a contest of technology and engineering, of ideas and audacity.

No major American car concern since the Duesenberg brothers in the 1920s had won a major contest in Europe, where racing marques were fueled by decades of innovation on twisty, unforgiving courses. American stock car racing—on oval speedways—was a different game, involving less sophisticated drivers and cars. Success could only be achieved by the marriage of brilliant design and steel-willed courage. It would require a greasy-fingered visionary to run the show, a team of the most skilled drivers in the world, and the swiftest racing sports car ever to hurtle down a road. All things of which, the optimistic Americans believed, could be purchased with the almighty dollar.

Henry Ford II's vision of his company as a Le Mans champion began as a marketing campaign, an investment he hoped would pay off at the cash register. In the end, it became something far more. Nationalism, glory, a quest to make history like no automotive magnate ever had—Henry II had discovered a way to conquer Europe in the unfolding era we now call globalism.

This is a work of nonfiction. All the events described in these pages actually occurred. The dialogue has been carefully reconstructed using countless interviews and contemporaneous accounts. Extensive notes on sources can be found in the endnotes.

PROLOGUE

June 11, 1955
6:24 P.M.
Le Mans, France

HIS NAME WAS Pierre Levegh and—sitting tight in the open cockpit of a silver Mercedes-Benz 300 SLR at the 24 Hours of Le Mans—the forty-nine-year-old Frenchman was about to become motor racing's most infamous man.

Two hours and twenty-four minutes into the race, Levegh found himself well behind the leaders. Coming out of a slight bend, he shifted from third to fourth gear and accelerated into a straight. The engine's exhaust note rose in pitch and volume, the wooden steering wheel throbbing in his hands. He was wearing goggles and earplugs, a United States Air Force fighter pilot helmet, blue trousers, light tennis shoes, and no seatbelt. Switching off with his teammate, the American John Fitch, he had nearly 22 hours of racing ahead of him, but already fatigue was gnawing at his discipline, his focus. He was alone in the cockpit; there was only himself and the car. Every intimation—a tug on the wheel, a second guess on the pedal—resulted in immediate response, fractions of a second gained or lost. On the dashboard, the tachometer needle arced across the gauge as Levegh accelerated past 135 mph, past 140.

Legions of fans crowded both sides of the pavement. A quarter of a million had come to this flat patch of central France to see what had been billed as a three-way battle for world domination between the silver Mercedes-Benzes, the green Jaguars, and the red Ferraris: the Germans, English, and Italians. No one had ever seen such beautiful cars travel so fast. Even at idle, they were the stuff of science fiction. Each ticket had a warning printed on it about the dangers of motor racing, but the spectators were otherwise occupied. The glamour of Le Mans was as intoxicating as the local wine.

To his left, Levegh saw the #6 Jaguar D-Type pull past him driven by Mike Hawthorn, an Englishman whom the French called The Butterfly because of the spotted bow tie he wore in the cockpit. Hawthorn was in a hurry. He was leading the race, setting a record pace on the 8.36-mile course that snaked through rural public roads. Levegh had just been lapped.

For the French driver, this was more than a race. It marked the culmination of more than thirty years of his life, three decades he'd spent chasing victory at Le Mans. He was nudging fifty and his future hinged on this one car, this last drive. As Hawthorn stretched his lead, Levegh saw the dream that had defined his life vanishing into thin air like the smoke from the Jaguar's exhaust pipes. It was all slipping away.

Levegh had had a vision years before—on May 26, 1923, the day the first Le Mans 24-hour *Grand Prix d'Endurance* was held. Two Frenchmen, Charles Faroux and Georges Durand, created the event to test the stamina and performance of cars and drivers, mapping out a roughly egg-shaped course through the countryside with twists and a backstretch for flat-out speed. A team of two Frenchmen won that first year and they walked away heroes. Levegh was there that day. He was seventeen years old. He promised himself that he would drive in the race one day. That he would win it.

Levegh began to study the craft of racing. He competed in his first Le Mans in 1938. Each year the race drew more fans, and each year the cars traveled faster. Like the spectators, Levegh sensed something magical about this race, something indescribably great. The rules were simple: a team of two men to each car, one man in the cockpit at a time. The car that completed the most laps over 24 hours won. Levegh nearly took the race once—in 1952. Leading with just one hour to go, he bungled a gear shift and blew his engine. By 1955, his prime was well past. They said he was washed up.

That spring, Alfred Neubauer, legendary manager of the Mercedes-Benz factory team, contacted Levegh. He wanted the Frenchman to have a car for the 1955 Le Mans. Mercedes officials knew that having a Frenchman on the team would be good public relations for the German company. There were a lot of car buyers in France who could remember the events of a decade earlier, when the Nazis leveled broad swaths of their nation. Meanwhile, Levegh knew that a spot on the Mercedes team—the world's most dominant in 1955—would make for the best shot he'd ever had. The 300 SLR was an open-cockpit racing car with two seats, trunk space, and headlights, all according to Le Mans rules. Underneath its lightweight magnesium skin lived a mechanical animal unlike any other. A 3.0-liter inline eight-cylinder engine dictated a long, bullet-like nose. The car featured a technology new at Le Mans called fuel injection. Top speed: in excess of 185 mph.

Levegh took the job.

In practice runs in the days before the race, however, he clocked slower times than the other Mercedes drivers, causing the portly team manager Neubauer to wonder whether the aging Frenchman had it in him. As the start of the race approached—4:00 P.M. on Saturday, June 11—Levegh paced the Mercedes pit with the look of a haunted man. He confided in his teammate about his fear of a particular part of the course—the narrow straight past the pits. "It is too narrow for these fast cars," Levegh said. "Each time I go by it I get a feeling of unease." Then: "A driver needs to feel comfortable, and I do not feel comfortable in this car."

"Levegh was going about with the face of a man in mortal terror," remembered journalist Jacques Ickx (father of the future star driver Jacky), who was covering the race. "It was the stuff of Greek tragedy. His pride, his immense obstinacy, would not let him admit that the car was beyond his capacity, that he should step down. All the time Mercedes believed that he would ask to be released.

They did not want to tell him that he was not up to it. So they waited for the resignation that never came."

At the wheel Levegh was stepping hard on the accelerator. Through a thin gauze of exhaust, he saw the grandstands at Le Mans rising in the distance, like an overflowing stadium sliced down the middle by a two-lane road. At the opening of the grandstands was the narrow straight past the pits. Directly in front of Levegh, "The Butterfly" Mike Hawthorn's Jaguar was pulling away. Hawthorn eased into the middle of the lane and lapped an Austin Healey with a British driver named Lance Macklin at the wheel.

Hawthorn, followed by Macklin, followed by Levegh at speed headed for the grandstands.

Suddenly the brake lights on Hawthorn's Jaguar flashed on. He was pulling to the right for the pit and braking hard, cutting off Macklin's Austin Healey. Levegh saw the Austin Healey's brake lights and smoke from under the rear tires as the car fishtailed. He had but a second to make his move. He eyed a 16-foot-wide alley on the narrow straight through which he could pass the Austin Healey on the left. He lifted his hand to signal the driver behind him to the obstruction, then turned the wheel, aiming to thread through. Traveling some 30 mph faster than the Austin Healey, he clipped its sloped rear.

Instantly the Mercedes was airborne. A 3,000-pound metal projectile with a tank of flammable liquid was 15 feet off the ground, rocketing at about 150 mph toward a crowd of spectators, with Levegh still hunched over the wheel. The car hit an embankment and exploded, hurtling fiery chunks of metal into the gathered mass. What was at one moment the social high point of the year—a party accented by the clink of wine glasses and the bellowing of sports car engines—became something unimaginably horrific. Dozens lay prostrate and bleeding. Fire raged; the car's magnesium body, made of a material similar to that used in early camera flashes, melted quickly into a thick soup of white-hot metal. Panic ensued. Those who were able to get on their feet fled and into that wave of foot traffic, photographers covering the race aimed and snapped, freezing shocked faces in black-and-white celluloid.

No one will ever know what went through Levegh's mind in that final second of his life. Madame Levegh? The fear of pain? Or was he seeing that checkered flag waving in the wind?

In the following days, readers all over the world opened their newspapers and absorbed the details of the tragedy in France. Photos resembled wartime images. The number of dead varied according to account, between seventy-seven and ninety-six.

But that wasn't the only strange part of the story. The race went on. Organizers believed that if they called it off, incoming roads would fill with traffic, blocking emergency vehicles. The Mercedes team pulled out in respect for the dead. This was a German car crashing into a crowd of predominantly French spectators; the Germans didn't want to start World War III. But the other competitors continued. Drivers dueled through the evening and into night. When the sun rose, they were at it still. At 4:00 P.M. on Sunday, nearly 22 hours after what was—and still is—the worst racing disaster in history, Mike Hawthorn took the checkered flag. He and his teammate Ivor Bueb had traveled 2,569.6 miles at an average speed of 107.07 mph, including pit stops and night driving. Record speed.

Following his win, a mob gathered around Hawthorn, who sat draped in victory flowers in the cockpit of his D-Type Jaguar, with its sharklike fin on the rear deck and feminine curves over the front wheel wells. Many were horrified by what they saw on the British driver's face. Framed between his flaxen hair and bow tie was the hint of a smile. They thought it scandalous, as many believed Hawthorn had caused Levegh's accident. But others in the crowd who knew Hawthorn couldn't blame him.

Racers came to Le Mans to become champions. And of the many champions crowned here, Hawthorn was the fastest of them all.

PART I

KINGS OF THE ROAD

THE DEUCE

> I will build a motor car for the great multitude.
>
> —HENRY FORD, 1909

HENRY FORD II opened his eyes. It was just before 8:00 A.M. on November 10, 1960. His toes hunted for his "HFII"-monogrammed slippers. He shaved left-handed, donned a fine-tailored suit over an "HFII"-monogrammed dress shirt, and stepped out of his seventy-five-room Grosse Pointe mansion into the Michigan sun.

He was forty-three years old, stood six feet tall, and weighed well over 200 pounds. Bright blue eyes blazed. Brown hair was side parted and slicked. Like many wealthy men of his generation, he had learned to show little emotion; his features had a stony quality, as if he was already turning into a bust that was going to sit in a museum. In his driveway, a black Lincoln limousine was idling—ample chrome, big V8 engine. The limo was not unlike the one Elvis Presley had just purchased, an update of the 1950 Lincoln that carted President Eisenhower around. Only this Lincoln had "HFII" discreetly painted on the door.

His commute took him west on Detroit's main artery, a highway named for his father, the Edsel Ford Expressway. He traveled through the city where, sixty-four years earlier, his grandfather first drove his "Quadricycle," a gas buggy with four bicycle wheels and a doorbell for a horn. In town, people liked to refer to Henry II as The Deuce, for he was the second Henry Ford, born the grandson and namesake of the world's richest man. But to his face, he was always "Mr. Ford." At the Michigan Avenue exit, the chauffeur veered off and pulled into the parking lot of the "Glass House" — Ford Motor Company world headquarters.

In his twelfth-floor corner office, Henry II took command of the world's second largest company. The Glass House, an aluminum and glass monolith he had constructed four years before, served as the brain center that maneuvered the tentacles of the man's increasingly global operation. Wherever there were roads, there were cars with his name on them. Ford Motor Company made Ford, Lincoln, and Mercury passenger cars; Ford trucks and tractors; Dearborn farm equipment; industrial engines; and military trucks. There was the Ford Motor Credit Company, the American Road Insurance Company, and the Ford Leasing Development Company.

Through his office window Henry II could see the stacks of the Rouge car plant billowing on the horizon. The Rouge was his grandfather's masterpiece, an industrial city that opened in 1920 with ninety-three buildings, 100 miles of internal railroad tracks, and sixteen locomotives. The stacks looked different now than they did in Henry II's early memories, having bled forty years of smoke into the ether.

Just before 10:30 A.M., Henry II told his secretary to summon an employee named Lee Iacocca. Minutes later, the thirty-six-year-old head of sales and marketing for Ford trucks and cars entered Henry II's corner office. Iacocca stood lanky at six feet, with olive-toned Italian-American features and a swoop of receding brown hair. His suit was a little flashier than the typical Dearborn getup. Not everyone in the Glass House knew how to pronounce the name Iacocca. He and Henry II had shaken hands but had never shared a conversation. Now here Iacocca was in Mr. Ford's corner office.

"It was like being summoned to see God," he later said.

"We like what you are doing, but we have something else for you," Henry II said that morning. After a pause: "How would you like to be a vice president and general manager of the Ford division?"

Iacocca gritted his teeth and smiled. As head of the Ford division, his job would be to bring out the new Ford cars—cars of the 1960s. The man who would soon be called "one of the greatest marketing geniuses since P. T. Barnum" got his first big break. Henry II followed with a little advice.

"If you want to be in this business and not lose your mind, you've got to be a little bold," he said. "You're going to make some mistakes, but go ahead."

When they shook hands, a relationship was born. Neither Henry II nor Iacocca knew at that moment that a revolution was about to grip their industry. A craving for speed was about to spread across the country like a contagion. Within months, it would begin to root itself in the public consciousness. For men who made cars, it would present extreme controversy, and a most magnificent opportunity.

This was Detroit—the Michigan city where the age of coal and steam had ended. It was a place where the grandeur of immeasurable wealth and the grit of an hourly wage communed. Detroit's car business grew so fast during the twentieth century, locals joked that the rise of the city's skyline could be measured in miles per hour.

In 1960, the Detroit companies were selling more than six million cars a year. The industry consumed 60 percent of the nation's synthetic rubber, and 46 percent of the lead. Unlike the men who founded American car companies early in the century—Henry Ford, Ransom Olds, the Dodge brothers—the auto men of the day didn't have to know how to design an engine. They did have to be good at math. "This is a nickel and dime business all the way through," Ford vice president Lewis Crusoe said. "A dime on a million units is $100,000. We'd practically cut your throat around here for a quarter." Companies were no longer run by the men whose names were on the cars. Except for one. In Detroit, Henry Ford II was royalty. "My name is on the building," he liked to say when his authority was questioned.

Henry II became a Detroit icon the day the nurses first swaddled him in the maternity ward. He was born on September 4, 1917. Any doubt about his destiny was put to rest when he was a toddler. In 1920, cradled in his famous grandfather's arms, he held a match to the coke and wood in Blast Furnace A at the River Rouge car plant. At three years old, Henry II breathed life into the largest manufacturing plant the world had ever seen while news photographers showered him with camera flashes.

He spent his early years in Grosse Pointe at his family's sprawling estate on Lake Shore Drive. His wealth isolated him; his childhood was anything but normal. The estate where he was raised had cement walls around it and a full-time security force lived on the property for fear of kidnapping plots. He had a little black button beside his bed that summoned his governess at any time of night, should he be thirsty. At Yale, Henry II couldn't pass engineering. He left school without a diploma after he was caught cheating on his final exam. When World War II broke out, he joined the military, as was expected of him, and was posted to the Great Lakes Naval Station.

During the years Henry II was away from home, the Ford empire, built up in the early part of the century to amass hundreds of millions of dollars, began to unravel. A cancer plagued the family, both literally and figuratively. The Fords had a dark family secret.

Throughout the 1920s and 1930s, Henry II's father Edsel had fought to modernize the company. He argued for a new breed of college-educated executive, and new cars that embodied style, personality, modernity. Henry I rebuffed him at every turn. The world was changing, but Henry wouldn't budge. "The customer can have any color he wants, as long as it's black," Henry Ford's famous saying went. General Motors, headquartered nearby in Flint, pressed hard to make its flagship Chevrolet America's brand of automobile. The national dailies ran banner headlines: "Ford-Chevrolet War Looms," "Ford-Chevrolet Battle for Supremacy."

Ignoring Edsel Ford's pleadings for a new model, Henry chose to fight it out with his obsolete Model T, lowering the price so much the roadster cost less per pound than a wheelbarrow. In 1924, two out of every three automobiles purchased in America were Model Ts. Two years later, Ford was being outsold two to one.

Edsel Ford's dream was to beat Chevrolet and make Ford America's car brand again. But he was powerless. On the eve of World War II, Ford held 22 percent of the market it had all but monopolized fifteen years before. Growing increasingly unstable in his elder years, Henry I installed a handgun-toting ex-convict named Harry Bennett in a top managerial position. Highly empowered, enforcing his will with his fists, Bennett harried and humiliated Edsel Ford while Henry I looked on.

As the company declined, Edsel grew ill and weak. He suffered vomiting episodes. He couldn't walk away from his job. He had an unerring sense of loyalty and duty and he had to think of his son: He had to protect Henry II's birthright. Edsel died at forty-nine in 1943 of stomach cancer, but those close to him believed he died of a broken heart, president of a family company in shambles, rejected by his father.

"He was a saint, just a saint," Henry II said of Edsel Ford on the day of his funeral. "He didn't have to die. They killed him!"

In September 1945, Henry I summoned his grandson to his estate, Fair Lane. The presidency of the company was open, and there was only one person who could rightfully fill it. Henry I offered Henry II the job, leaving the young man at a crossroads. He could walk away and live a life of leisure forever, wealthy beyond most Americans' ability to fathom. Or he could take upon himself this family legacy with all its fame and warts. Only recently had he come to understand all Edsel Ford had endured. *This thing killed my father,* Henry II thought to himself. *I'll be damned if I'm going to let it kill me.*

Henry II accepted the position—on one condition. That condition set the tone for the company's next forty years: "I'll take it only if I have completely free hand to make any change I want."

Days later, at age twenty-eight, Henry II moved into his father's office and placed a photo of Edsel Ford behind him, so his father was looking over his shoulder. A young man with a face full of baby fat, a high-pitched voice, and almost no business experience to speak of took the wheel of a corporation spinning out of control. Ford Motor Company was hemorrhaging millions of dollars every month. It was impossible to give an exact number because there was no accounting system. "Can you believe it?" Henry II later remembered. "In one department they figured their costs by weighing the pile of invoices on a scale." During World War II, the company had built 86,865 aircraft, another 57,851 aero engines, and 4,291 invasion gliders, and the task of retooling the factories to produce road cars was at hand. There were forty-eight plants in twenty-three countries.

All signs pointed to disaster. One Ford executive referred to Henry II as "the fat young man walking around with a notebook in his hand." Henry II chose his rallying cry, two words he had printed on signs and plastered all over the Rouge.

"Beat Chevrolet."

Henry II's first task was to bring in the college-educated executives his father had fought for. Soon the "Whiz Kids" were pacing the hallways at headquarters, a brilliant ten-man team plucked directly from the Army Air Force who, during World War II, had been in charge of the statistical and logistical management of the waging of war. Among them was Robert S. McNamara, future Secretary of Defense in the Kennedy and Johnson administrations. Recruiters visited fifty universities aiming to hire the top engineering student at each. One graduate student at Princeton named Lido Anthony Iacocca signed on as a $185-a-month trainee.

"Look, we're rebuilding an empire here," the company's legal chief told a lawyer he was interviewing. "Something like this will never happen again in American business."

Henry II gambled his family legacy on one car. On June 8, 1948, he beat Chevrolet to produce the first all-new postwar model, and personally unveiled it at New York's Waldorf-Astoria Hotel. He spent nearly $100 million on the launch, an unheard of sum at the time. It was a slinky Ford with a snub nose and enough interior room that a man could drive with his hat on. Tens of thousands turned up to see the new car and the new Henry Ford. Henry II strode the carpeted floors and smiled nervously. He did not have the appearance of greatness, not a hint of charisma. Among the mingling crowd was the recently widowed Mrs. Henry Ford senior. She could still recall attending her first automobile unveiling, her late husband's Model T.

"It was more than 40 years ago," she told a reporter, "and I thought it was the finest exhibition man could create. Who could have imagined then anything as splendid as what I see here?"

The papers called the new Ford Henry II's "instrument of conquest." More than one hundred thousand orders came in the first thirty days.

Chevrolet countered in 1950 with the top-of-the-line Bel Air. Ford answered with the Crestliner. Chevy launched the Corvette at its 1953 Motorama and Ford came back with the Thunderbird for 1955. As the two giants stood toe to toe, more Americans chose sides, shelling out for new vehicles. The 1950s saw a crystallization of the middle class—Detroit's prime customer. Suburban neighborhoods reachable only by car blossomed over the forty-eight states. All the elements conspired to lead the industry into a Renaissance unlike any in history. American cars with big engines and bright paint were symbols of prosperity and supremacy.

By 1955, Americans were buying up vehicles at a record pace, and Ford and Chevrolet were running neck and neck. Henry II had engineered what many were calling "the greatest corporate comeback in history." Dearborn was electrified.

On February 21 of that year, President Eisenhower spoke out in favor of the new Federal-Aid Highway Act, which would pump $100 billion into creating a nationwide interstate highway system, the greatest system of government-controlled free highways in the world. That same week, Henry II picked up an issue of *Life* magazine and saw his face staring back at him. It was the face of a thirty-eight-year-old who had grown accustomed to his position as one of the most influential men in America. His face now radiated power, his eyes communicating a distinct fearlessness. "The fight between Ford and Chevrolet has given the public a spectacle combining some of the best features of the World Series, a heavyweight championship fight, and a national election," the *Life* story read. "Ford's fight to regain the leadership it lost in the 1930s, resulting in the biggest industrial struggle of our times, is an extension of one of the most remarkable human stories of any time."

The Deuce had arrived.

It was this world of money and cutthroat competition that Lee Iacocca stepped into in the winter of 1960. Iacocca's promotion to top man at the Ford division occurred the same week that John F. Kennedy was voted into the Oval Office. Inside Iacocca's nerves slithered, but his outer shell was impenetrable. His colleagues saw a salesman through and through. After listening to Lee talk about a car, one once said, "I didn't know whether to drive it home or make love to it." Iacocca lunched with Mr. Ford in the Glass House's penthouse dining room—hamburgers with ketchup served by waiters in white coats. There were meetings in Henry II's office, and company parties that went on and on. Henry II could mix scotch, wine, and cigarettes all night and never seemed to have a hangover.

Iacocca embodied the other end of the American dream. He was the son of Ellis Island immigrants. Raised in Allentown, Pennsylvania, he was the first top man at the Ford division who had real dealership experience, selling on the floor, and he hadn't shed the bootstrap urgency. Business was personal. He had rougher edges than his associates. He could work the phrase "son of a bitch" into every other sentence. He had dreamed of working at Ford since he was in high school. The day a recruiter from Dearborn showed up at Lehigh in a big Lincoln, and Iacocca smelled the leather interior, his resolve had only strengthened.

During his first months on the new job, Iacocca took a hard look at the American car buyer, not just in present day but what that customer was about to become. He had the benefit of a new tool feeding him information. Computers had arrived in offices. They could make sense of market research in a way a person never could. They answered the question: Who bought what? Where? And when? Sitting in Henry II's office one day, Iacocca began to map out his vision of 1960s America. He clutched his Ignacio Haya cigar like a stage prop, and he armed himself with numbers. He was very good at math.

"Next year there will be a million more 16-year-olds than there are this year," Iacocca said. "By 1970, there will be over 50 percent more people in the 20 to 28 year old group than there are today."

The key to the future, Iacocca reasoned, was the youth market. The war-baby generation was coming of age. Soon the kids would be filing into government offices to get their driver's licenses, and unlike their parents' generation, a huge portion of them would have the money to buy an automobile. Trends come and go, but one thing never changes: The primary goal of every sixteen-year-old is to own a car. Iacocca called the new generation "the buyingest age group in history." If Ford could get the kids when they were young, he argued, Ford could get them for life.

One day early that spring, Iacocca sat down to lunch at the Detroit Athletic Club, a well-appointed watering hole where the walls were lined with portraits of Detroit's pioneers. The auto men who lunched at the club discussed their cars the way others spoke of their children. That afternoon Iacocca was passed a photograph of a new Chevrolet called the Monza, named after Italy's famed Grand Prix circuit. Chevrolet had taken its staid compact, the Corvair, and added bucket seats, Euro-style trim, a four-speed stick, and enough engine to power the thing up to 125 mph.

The muscles in Iacocca's jaw tightened. *Those sons of bitches at General Motors saw the same vision of 1960s America that he did.* The race was on.

Iacocca gathered a team of admen, engineers, and designers and began to outline a new theme for Ford. "What we need is a campaign, a philosophy that is young at heart," he told them. The team went to work in a large windowless room at the Ford office called "the tomb," where Iacocca enforced security so rigid the garbage was burned under supervision. Weekly late-night meetings were held at a nearby motel called the Fairlane Inn. Iacocca's underlings began to understand the depth of his ambition. He was proof that you couldn't have enough balls in this town.

Iacocca's vision included a new model, and a marketing theme to sell it, a way to make a statement: Ford cars were about performance, style, adventure. He knew America was ripe for something different. The numbers were there in the market research, and he could feel it in his gut. Together with his chief engineer Donald Frey, he approached the styling department. Within days, a secret prototype was in the works. The original name was Torino, after the Italian city Turin. But it would one day soon become a car called Mustang.

"Put in class for the mass," Iacocca told his designers. "If we're right, this will make the Model A look like nothing."

On Monday morning, February 27, 1961, automobile executives throughout the Detroit suburbs awoke to the sound of newspapers slapping onto their stoops. The *Detroit News* was an international paper, but it was also a car business trade journal and gossip sheet. That Monday was a slow news day. "Liz Taylor Ill." "Venus Probe Monitored by Soviet Mystery Ships." On the sports page the daily featured in-depth coverage of the Daytona 500. On Sunday, Marvin Panch had cruised to victory in a Pontiac in the fastest 500 in history—an average of 149.6 mph around Florida's famous 2.5-mile thunderdome. In columnist Doc Greene's commentary, readers discovered some interesting information:

> In European racing, victory can be translated immediately into sales. Buyers over there operate on the rather simple theory that if, for example, in the 24-hours endurance race at Le Mans, five Ferraris finish ahead of the rest of the pack under such grueling circumstances—it's the best car and you ought to buy it. "It has worked the same way for us," remarked [Detroit Pontiac dealer Bill

> Packer, who sponsored the winning car]. "Back in 1957 when Bunkie Knudsen took over the division, a Pontiac was a good car all right but it had a reputation for being an old woman's auto. Great for grandmas. Then we started dominating stock car racing. We went way up in sales in just a couple of years.

Stock car racing was not supposed to matter in Detroit. There was a ban prohibiting automakers from spending a dollar on racing. In the late 1950s, American carmakers came under pressure for encouraging drivers through advertising to put horsepower to the pavement on public roads. Ford, General Motors, and Chrysler were involved in a "Horsepower War," each trying to outdo the other with bigger engines in hopes of capturing customers. Stockpiling power in cubic inches—it was Detroit's answer to the Cold War. The companies marketed their cars by entering them in a series of Sunday skirmishes called National Association for Stock Car Automobile Racing—NASCAR.

The relationship between the battle for speed and the battle for market share didn't sit well in Washington. In 1957, Congress forced the Automobile Manufacturers Association to draw up a "Safety Resolution," a treaty by which Detroit automakers agreed not to advertise "the specific engine size, torque, horsepower, or ability to accelerate or perform, in any contest that suggests speed." "Safety" and "economy" became the buzzwords of the day.

And yet, Doc Greene's column in the *Detroit News* spoke of high-powered automobile executives showing up in bunches at the Daytona 500 in 1961. "The executives you see just happened to be passing through on the way to Texas or Dubuque or somewhere and you didn't see them at all. And the things they say are off the record." One "wheel within a wheel" was quoted: "Racing without cooperation from the parent company is like having a baby without an umbilical cord. Almost impossible."

That summer General Motors' investment in racing became the worst-kept secret in Detroit. Chevrolet was funding a racing campaign under the guise of a marine engine program. Pontiac had its own secret program.

"These guys are cheating," Henry II told Iacocca. "We have to do something."

Henry II refused to spend money on racing, fearing bad publicity. He was president of the Automobile Manufacturers Association. Enforcing the Safety Resolution fell to his authority.

By the end of 1961, Iacocca's first year on the job, Pontiac and Chevrolet had won forty-one out of fifty-two NASCAR races. Over that same year, General Motors' share of the market began to skyrocket. Pontiac reported its highest new model sales in its thirty-six-year history. In April, Chevrolets won at Richmond, Columbia, Greenville, and Winston-Salem. Over that same thirty-one days, Chevrolet set an all-time single-month sales record. "Not since the heyday of Henry Ford's Tin Lizzie has any auto line so completely dominated the market," *Newsweek* reported. "One of every three cars that Americans drove out of showrooms last week was a Chevrolet."

Anxiety began to cloud the hallways of the Glass House. Since Henry II took over the company, he'd never seen such an immediate and critical shift in market share. Ford executives saw stories in their newspapers about young people who were taking to the streets and marching, demanding legal drag strips in their hometowns. Kids wanted speed, and General Motors made the hottest engines money could buy.

On April 27, Henry II penned a letter to John F. Gordon, president of General Motors. "On a short-range basis we intend to develop those high-performance components for the Ford car that are presently offered by Chevrolet and Pontiac," he wrote. "We believe this action is mandatory to assure the competitive position of our products. On a long-range basis we are most anxious to work out a more satisfactory agreement with other members of the industry."

Henry II never received a reply.

In the first third of May 1962, GM's slice of the pie hit 61.6 percent, up from 49 percent a year before. Not in forty years had one company cut out such a large chunk. Attorney General Robert Kennedy had a grand jury looking into antitrust laws. From his fifth-floor office at Ford division headquarters, Iacocca saw Ford spiraling under his watch. He had worked his whole adult life to get this job. He urged his boss to scrap the Safety Resolution—out of necessity.

"Now I don't want to imply that we were building old ladies' cars," Iacocca later said. "But something had to be done. I had only one thing in mind. We had to beat hell out of everybody."

It was an easy sell. Henry II saw not a company but a family legacy in peril. An attack on Ford was an attack on the Ford family. How many times had Henry II heard the story as a child, how the family legacy was founded? One cloudy Thursday many years before, on October 10, 1901, an unknown tinkerer named Henry Ford stepped into a machine he had built, at a racetrack in Grosse Pointe. A crowd of six thousand had gathered to see a sweepstakes race between this unknown tinkerer and Alexander Winton, holder of the world track speed record and the most famous name in speed of his day. Henry Ford didn't believe in racing; he thought it irresponsible. "But as others were doing it, I too had to do it," Henry I later described this moment. "If an automobile was going to be known for speed, then I was going to make an automobile that would be known wherever speed was known."

Henry I won the race. He gambled his life on it and prevailed, and this victory put a new carmaker named Henry Ford on the map. What would history be if Crazy Henry had come in second?

On June 11, 1962, Henry II released a statement—six paragraphs with his signature at the bottom—regarding Detroit's Safety Resolution. "Accordingly," he told reporters that day, "we are withdrawing from it." Anticipating criticism from Washington, he asserted that the company would "continue with unabated vigor our efforts to design, engineer and build safety into our products." Off the record he was more to the point.

"We're going in with both feet."

It was, in the words of one Detroit reporter, "the biggest automotive scoop in years," and reaction was immediate. Representative

Oren Harris, Democrat of Arkansas, promised an investigation by the House Health and Safety Subcommittee. Editorials probed Henry II's motivations. "Detroit's romance with racing has always been a strange one," the *New York Times* stated. "Cynics say it is a romance of desperation, pursued only when other approaches don't bring satisfaction in the salesroom."

Henry II knew that, if Ford was going to lead the charge, the other companies would likely follow. Competition would move from the boardrooms and showrooms to the track. Losing was out of the question.

2

IL COMMENDATORE

> I am convinced, that when a man tells a woman he loves her, he only means that he desires her; and that the only total love in this world is that of a father for his son.
>
> —Enzo Ferrari, 1963

Through his windshield, the city of Modena, Italy, came into focus: rows of stone buildings, telegraph wires, the towering spire of the Duomo poking at an iron-gray sky. The maze of teeming ancient streets pulled him in as if by a force of gravity.

His name was Enzo Ferrari and he was fifty-eight years old. He made the famous cars that bore his name, and yet he drove a Fiat the 11 miles home from his factory. Once he arrived in the city, the streets clogged with traffic: Fiats like his own, Lancias, an occasional Ford. Bumpers scraped and scooters swarmed. It was a June evening in 1956. The setting sun reddened the horizon but the early-summer heat was still baking Modena.

Turning onto the Viale Trento e Trieste, his Fiat lurched to a stop on a corner in front of number eleven. This was his home—a two-story painted stone structure with two Shell gas pumps out front skirting the road. On the first floor were the garages, big enough for a half dozen automobiles, where his customers came to have their Ferrari cars serviced. He kept an office on one side, a small chamber with a desk, a well-worn brown leather chair, and a trophy-lined bookcase.

He climbed a stairwell to the second floor, where he resided with his wife Laura and their only child, a son named Alfredo—Dino for short. The walls embraced Ferrari like the fabric of his old suit. He had not strayed from these rooms for a single night in perhaps more than a decade. His wife Laura's sad face at the end of the day was as familiar to him as the furniture. They seemed to have a relationship that neither enjoyed but neither could do without.

Was there any news?

There was not, she informed him. The doctor had come and the doctor had gone. What was there to say?

Ferrari walked into his son's room. Dino's six-foot frame filled his bed. He was twenty-four and he'd been born into this home. The older man's skin had leathered and his dark hair had grayed, but the two still shared the same features: prominent forehead, aquiline nose, hair bristling back as if it were being blown by a steady breeze. They were unmistakably father and son. Ferrari eased himself into the chair that had become a fixture by Dino's bedside and time slowed to a crawl.

Conversation started where it had ended the evening before. Ferrari updated Dino on what was happening at the factory, whom had visited, the business of the day. It was June, which meant Le Mans approached, the most important race of the year. In the frenzy to ready the cars the daylight hours moved at breakneck speed. There was always disagreement. Bazzi had said this . . . Jano had said that . . . In the racing shop, voices could sometimes drown out the rattle of a sick engine. Throughout his years Dino Ferrari had asked his father about the *cucciolo,* the "puppy." He liked to call his father's latest creation the puppy. Ferrari was building the 625 LM to race at Le Mans in 1956 and he detailed the progress.

Inline four cylinder, 2.5 liters, two Weber carburetors . . .

By the bed, Ferrari kept a notebook. Inside he had scribbled charts and graphs tracking the number of calories his son had consumed on a given day, his diuresis, the presence of albumin in his urine and of urea in his blood. Visits from a doctor enabled Ferrari to update the figures frequently. Like an engine on a test bed, Dino's numbers measured vitality and strength.

Ferrari had always believed that the young man's problems could be solved. Even in those late hours, he believed his son could be saved.

Dino Ferrari had lived his entire life in a world of experimental automobiles. From his earliest days he could be seen following his father around in the garages below their home and at the local proving ground, the Modena Autodrome. He was in his early teens when he began suffering from strange maladies. Doctors eventually diagnosed him with muscular dystrophy,* a crippling disease with no cure that eats away at the skeletal muscles.

Ferrari groomed his son. Dino earned his engineering diploma at a technical institute in Modena, and unlike his father he learned to speak English. He took an office near his father's and began to assume his role as the aging boss's only son.

"Papa," he said when he saw Ferrari upset about a business matter. "Don't let it get you down. Things always right themselves if you only give them time."

By the end of 1955, Dino's twenty-four-year-old legs had grown so stiff he had trouble walking, and his kidneys began to fail. He was confined to his bed.

Frequent visitors came to sit with Dino—his best friend Sergio, and drivers on Ferrari's Grand Prix team. The filmmaker and avid Ferrari customer Roberto Rossellini came bearing books, and he sat by Dino's bed for hours. Dino's most important visitor came at night. After sunset Enzo Ferrari returned home from the factory with a man named Vittorio Jano. Jano was in his mid-sixties and always dressed in his trademark three-piece wool suit and bowler hat. He was the most famous engineer in Italy. In the 1930s, Jano had designed the Alfa Romeo P3 *monoposto,* the first true single-seat racing car. In a sport that was as much about mechanical wizardry as it was about the skill of drivers, Jano's cars had made

* Historians have debated the nature of Dino Ferrari's disease, arguing everything from leukemia to syphilis passed on by his mother at birth. Most agree he was afflicted with muscular dystrophy.

Italy the nation to beat. His contribution to Ferrari's rise was immeasurable. Ferrari could still recall the day, thirty-three years before, when he first climbed the stairs to Jano's home and knocked on the door.

In Dino's room, the three men resolved to design a 1.5-liter racing engine. The challenge: to achieve the perfect equilibrium between power and economy of motion. Racing historians would forever debate exactly what occurred in that room. Ferrari would claim that his son was achieving greatness on his deathbed, though some argued Jano's genius was at work.

As the engine took shape on paper, and the young man's condition deteriorated, a macabre irony became apparent. For Enzo Ferrari, the internal-combustion engine was a symbol of life. It had revolutionized society. He had watched it all happen during his lifetime. He spoke of automobiles as if they were animate. Cars possessed unique behaviors. They breathed through carburetors. They were skinned with metal. "Ferrari's aim," he once told a reporter, addressing himself in the third person, "is to perfect an ideal, to transform inert raw material into a living machine." The engine of a car was both heart and soul. Ferrari engineer Luigi Bazzi had called its rumble "the heartbeat of the creature."

Spread out around the Ferrari home, the city of Modena was going about its routine. A thriving colony since the days of Caesar, Modena (MO-deh-na) was settled on the Via Emilia, a strategic road that cut a path straight to Rome. Modena's population in 1955 was 150,000, roughly the size of Yonkers, New York, today. It was a city of cobblestone alleys and paved streets winding from the Piazza Grande, where a statue of the Madonna held court next to the Duomo, built in the twelfth century, its tower now slightly tilted.

Italians singled out Modena for its *zampone* (stuffed pig's feet), a sparkling red wine called Lambrusco, balsamic vinegar, and a lousy soccer team called *Il Canarini*—the Canaries. Its true fame, however, was its craftsmen. "Absurdly gifted artisans abound," the historian Griffith Borgeson described the city, "so that you can have almost anything made, made surpassingly well, and so cheaply that you must never get used to the miracle. It's an incredible place, where master pattern makers are a dime a dozen and skilled metal workers of every kind seem to surge out of the black humus."

As in all Italian cities and towns, the Modenese held beauty in great esteem. Everywhere it was evident, from the Renaissance architecture to the label on a bottle of Lambrusco. Modena's metal workers possessed a talent for mechanics as well, an instinctive ability to design and mold moving parts. In this city, an old-world aesthetic was joined by modernity's defining ambition: to harness power. "It is my opinion," Ferrari once wrote, "that there are innate gifts that are a peculiarity of certain regions and that, transferred into industry, these propensities may at times acquire an exceptional importance . . . In Modena, where I was born and set up my own works, there is a species of psychosis for racing cars."

The city's streets and alleys were lined with coach-building shops and foundries. Maserati, among the world's most storied racing marques, had its headquarters here on the Via Ciro Menotti. The famous Weber carburetors were built in nearby Bologna, also the home of motorcycle fabricator Ducati. Just a few miles down the road from Modena in Sant'Agata Bolognese, a prominent tractor manufacturer would soon unveil a phenomenal first car, his name Ferruccio Lamborghini. It was Enzo Ferrari, however, who was emerging as the region's most revered mechanical impresario.

Ferrari was a metal worker's son; his name came from the Italian word *ferro,* meaning iron. He could describe himself as neither a designer nor an engineer, but rather "an agitator of men." Mornings found him exiting his apartment wearing a plain baggy suit, suspenders holding up his pants. On his wrist was an old chronometer with the black Prancing Horse—the Ferrari logo—in the center of the dial. He visited his barber Antonio for a shave, then drove a Fiat 11 miles to the rural village of Maranello, where an Alsatian named Dick greeted him at his factory gate, tail wagging. The factory porter Seidenari—whom Ferrari's wife called "Sei di

Denari" ("You have money")—snapped to attention in the porter's lodge. The boss's office was there by the gate, so he knew who was coming and who was going.

He worked seven days a week, twelve to sixteen hours a day, holidays included. At night, he returned to Modena. He felt "extremely emotionally attached" to his city. Except for his daily drives to Maranello, he refused to leave Modena for almost any reason. He did not attend races—not even the Italian Grand Prix at the Monza Autodrome, one of the world's most famous circuits, a two-hour drive from his home.

The factory produced just a few cars each week, of three different kinds: two-seat racing sports cars of the Le Mans type, missile-shaped single-seat Grand Prix cars (increasingly being called Formula One at this time), and Grand Touring cars (customer road cars). Paying for it all was a rarified group of clients, who commissioned their cars as if they were pieces of art and paid Enzo Ferrari extraordinary amounts for them: the president of Argentina Juan Domingo Perón, the Shah of Iran Reza Pahlavi, the playboy Porfirio Rubirosa and his mistress Zsa Zsa Gabor. When they came to Modena to take possession of their vehicles, they always stayed a little longer to meet Ferrari. They addressed him by the title the Italian government had given him in 1927—*Commendatore.*

To the journalists who sought to define him in their articles, Ferrari was a riddle: A man who built racing cars, but refused to attend races. Who worked tirelessly to perfect state-of-the-art machines, yet feared elevators. They called him The Magician of Maranello, a "speed-bewitched recluse." But here in his home city, where he had grown up, there was no mystery. He looked and dressed like any other man sitting in the cafes of the Piazza Grande. He was a Modenese, a *paesano.*

On June 30, 1956, a priest arrived at 11 Viale Trento e Trieste. Dino was dying, and the priest came to sit with his parents through the final hours. In the chair by Dino's bedside, Ferrari sat with his son. They talked, and from their conversation, Ferrari learned, as he would later describe, "what life means to a young man who is leaving it."

When Dino had taken his last breath, Ferrari opened the notebook that he had filled with charts and graphs and wrote one final sentence: "The match is lost."

"I have lost my son," he said.

"Let us say a prayer for our Dino, who has left us," the priest said.

"Since I took First Communion as a child I have forgotten the prayers that so many people say every day," Ferrari responded. "The only thing I can say is: God, help me to be a good man."

The following day, drivers in Ferrari's red livery competed in the French Grand Prix wearing white armbands in Dino's honor. The British driver Peter Collins won the race for the Ferrari factory team, traveling 121.16 mph over 314.5 miles—the fastest average speed over any European circuit in history. When the phone call arrived, Ferrari waved it away. He claimed that racing no longer had meaning for him and announced that he would give it up forever. No one believed him.

Dino's body was eased into the Ferrari family crypt at the San Cataldo cemetery in Modena. Enzo Ferrari slipped on a pair of dark sunglasses that became as much a part of his face as the nose they sat on. The office Dino kept was left intact; his possessions—his appointment book, his pen—remained exactly where they were when his fingers last touched them. In his own office at the factory, Ferrari mounted a portrait of his son above a sconce, like a shrine. Each morning Ferrari made a ritual of brooding at Dino's tomb.

The man was now fifty-eight years old, with no legitimate heir to carry on the legacy he had dedicated his life to creating. He believed that he had lost everything that mattered to him, with the exception of one thing: his cars. He would deny them nothing.

Ferrari could remember the day he was seduced by automobiles. He was eleven years old. The year was 1909.

He lived with his parents and brother in a small flat above his father's metal shop on the outskirts of Modena, and every morning he awoke to the sound of hammers clanging below. One day he rose from bed and set out across the rail tracks adjacent to his home. He hiked two miles alone. In the countryside the Modena Automobile Association had organized a race called the *Record del Miglio*. A group of gentleman drivers were going to attempt to break the mile speed record.

Donkeys far outnumbered automobiles on Modena's streets in 1909. Motorcars were objects of curiosity, and a chance to see how fast they could go lured a bustling crowd. Enzo took in the scene: Mechanics paced next to piles of tires and fuel drums. Time keepers sat at a table near a banner with FINISH written on it by hand. The silence gave warning and then it appeared: the thunderous racing car. A man named Da Zara set the best time in the flying mile that day: 87.148 mph.

World War I derailed Ferrari's ambitions. The war ravaged Italy, destroying its economy and infrastructure. And yet, the war accelerated the innovation of automobiles and airplanes. Sophisticated new machinery resulted, as well as a generation of men accustomed to speed and danger. They had not forgotten their lost brothers. Bitterness lingered, and in the 1920s car races served as symbolic warfare. Cars raced in national colors: red for Italy, blue for France, green for England, yellow for Belgium, white (later silver) for Germany.

The war left Ferrari penniless, his father and brother dead. He was relatively uneducated, having sat through four years of elementary school and three of trade school. But he possessed a valuable talent—a knack for fixing things.

At age twenty-three, he joined Alfa Romeo as a test driver, mechanic, and competitor. He earned his first victory on June 17, 1923, at a race in Ravenna. While he looked out from the podium, a man pushed through the crowd and introduced himself as the Count Enrico Baracca, father of the Italian war hero Francesco Baracca, a Modenese who'd shot down thirty-four enemy planes before he was killed in 1919. The ace pilot flew with a black Prancing Horse, the symbol of his squadron, painted on his plane's fuselage. It was Baracca's mother who told the young racing driver: "Ferrari, why don't you put my son's *Cavallino Rampante* on your car? It will bring you luck."

Ferrari laid this Prancing Horse symbol against a yellow shield, the color of Modena. This badge would one day become one of the most recognizable brand symbols in the world.

In 1929, Ferrari founded the Scuderia Ferrari, a private team that served as Alfa Romeo's racing arm. (Literally, *scuderia* means stable, as in a stable of talent, or a team.) Headquarters was a two-story stone building on Viale Trento e Trieste. Enzo moved into the apartment above with his new wife, a peasant woman named Laura. He won his last race in 1931. The following January, Laura gave birth to a boy the couple called Dino. Ferrari declared that he would never race again. He had an heir, and he devoted himself to creating a legacy that would live beyond him.

Ferrari believed the winning formula was 50 percent car, 50 percent driver. As for cars, he had armed himself with a fleet of red Alfa Romeos designed by the man who had emerged as Italy's foremost mechanical maestro, Vittorio Jano. What the *scuderia* needed was a star at the wheel.

"The Flying Mantuan" Tazio Nuvolari was a man of few words, but his jutting jaw and cold eyes screamed of valor and defiance. Nuvolari combined precise skill with what Ferrari called "near superhuman courage." He was, by many estimates, the greatest racing driver in the world—ever. Ferrari negotiated a deal that paid the Mantuan almost a third of the *scuderia*'s income. It would prove a stroke of genius.

Once, during a prerace practice, Ferrari passengered with Nuvolari. "At the first bend," he later remembered, "I had the clear sensation that we would end up in a ditch; I felt myself stiffen as I waited for the crunch. Instead, we found ourselves on the next straight with the car in a perfect position. I looked at Nuvolari. His rugged face was calm, just as it always was, and certainly not the face of someone who had just escaped a hair-raising spin. I had the same sensation in the second bend. By the fourth or fifth bend, I began to understand. I had noticed that through the entire bend Tazio did not lift his foot from the accelerator, and that, in fact, it was flat on the floor."

In Jano and Nuvolari, Ferrari had prototypes: an engineering mastermind and a champion pilot.* The duo powered the Scuderia Ferrari into the limelight in Italy. During these years "the agitator of men" studied the psychology of winning. Certain principles were self-evident.

1. Competition is the impetus for innovation. The fiercer the competition, the faster cars will go.

2. There is in some men a need to achieve greatness. When matched with talent, this necessity can turn humans into demigods.

3. A man who is willing to die at the wheel is always likely to beat a man in a faster car—if he can survive until the end of the race.

During the 1930s, German chancellor Adolf Hitler began to understand the symbolism behind the Grand Prix car. He offered huge sums of reichsmarks to any German firm that could produce a successful racer. A fleet of "silver arrows" resulted—massive machines from Mercedes and Auto Union. To reduce weight, the Germans stripped the cars of paint so the metal gleamed. Beginning in 1934, the Germans dominated racing, to Hitler's delight.

In July 1935, Nuvolari and a caravan from the Scuderia Ferrari left Modena and traveled north to the German Grand Prix at the Nürburgring, one of the first purpose-built closed racing circuits, 14.2 miles of mountainous twists. Nazi storm troopers escorted the cars to the starting grid as two hundred fifty thousand spectators

* Italians used the word *pilota,* or pilot, to denote a racing driver. The word "driver" meant chauffeur.

looked on. Everyone believed the German cars to be invincible. The Führer himself was most certainly listening in by radio. The silver arrows lapped at outrageous speeds with swastikas painted on their bodies. The red Alfas had on their fuselages the black Prancing Horse against a yellow shield.

Manfred Von Brauchitsch, nephew of a high-ranking Nazi official, led the race in a Mercedes-Benz. But Nuvolari hunted him down. As the racers tore into the last lap and the crowds looked on amazed, a voice piped over the loudspeakers:

"Brauchitsch has burst a tire! Nuvolari has passed him! Brauchitsch is trying to catch up on a flat tire!"

At home in Modena, Ferrari's phone rang. He heard the news: Nuvolari had won the German Grand Prix. In Italy, racing was a passion that joined all classes in every region; entire cities erupted in celebration. Ferrari's reputation was cemented. The Prancing Horse came to represent not just a man and his cars, but a nation.

Ferrari built his factory in Maranello during World War II to produce machine tools. Surrounded by empty fields, it consisted of three rows of shed-like buildings shaped in a triangle with a cobblestone courtyard. An eerie calm marked those first months. It was interrupted in November 1944 by American bombers, who pummeled Ferrari's factory. By the summer of 1945, the violence had begun to subside and the factory was patched up. Plans for the first Ferrari car were in the works.

That winter, Ferrari received an unexpected guest. Luigi Chinetti was a towering figure in European racing circles. He was a two-time Le Mans champion, driving Alfa Romeos on both occasions. Chinetti's path had taken an unexpected turn. Strongly antifascist, he'd escaped Europe on the eve of the war. He'd signed on as chief technician in the French champion René Dreyfus's expedition to Indianapolis. Dreyfus, an officer in the French army, competed in a Maserati at the Indy 500 on May 30, 1940, with Chinetti working the pit. That same week, 350,000 troops were evacuated from Dunkirk. Chinetti defected and ended up in New York, where he was employed at J. S. Inskip's Rolls-Royce dealership.

Now he was back in Europe to see what was left of his past. He found his friend Ferrari shivering in an overcoat in the old *scuderia* office. The room was ice cold and dimly lit, a single bulb hanging from a chord from the ceiling. Chinetti was an American citizen now, and he described what he had found in the new world. Roads were filled with big Detroit cars, but the sports car did not exist. There was no such thing. Chinetti had an idea: Ferrari should build cars and sell them in America.

"The future is here, Ferrari," he said of his adopted country. "You must believe that here sports cars will be a gold mine. There is hunger for motor sport. The market is virgin. There is plenty of money. The potential is immense."

Ferrari agreed—if Chinetti was willing to put up the money. Chinetti returned to New York and set up shop in a small garage on West 49th Street. Without a car to sell, he declared Ferrari open for business.

It took Ferrari nearly two years to build the first car. In postwar Europe, electricity was a luxury. Fuel and manpower were in short supply. Ferrari chose a V12 for the layout of his first engine. "All we wanted to do was to build a conventional engine," he later recalled, "only one that would be outstanding." In truth, the engine was hardly conventional. Its twelve cylinders were the diameter of silver dollars, designed to rev at very high speeds. The first Ferrari—the 125—debuted in a race at Piacenza on May 11, 1947. Its first victory came two weeks later at the Rome Grand Prix.

As America's Marshall Plan dollars flooded Europe's economy, Ferrari hurled himself into the postwar racing scene. In 1948, Ferrari cars won the 12 Hours of Paris, the Mille Miglia, the Targa Florio. A year later, France announced that it would hold the most prestigious of all races again after a nine-year hiatus—the 24 Hours of Le Mans. The first postwar Le Mans was won by Luigi

Chinetti and his teammate, Lord Selsdon of England, in a Ferrari 166 MM, a roadster nicknamed *barchetta* ("little boat"). In a truly heroic display of endurance, Chinetti drove all but a few laps over the twenty-four hours (his teammate Selsdon is rumored to have been drunk). The win triggered an instant demand for Ferrari cars across the continent.

The only cars Ferrari had to sell were battle-scarred racers. But by 1949, he began offering customers touring cars in small numbers. Chinetti made his first sales. The first Ferrari arrived in the United States in June 1949; the buyer was Briggs Cunningham of Connecticut, the renowned yachting champion and soon-to-be racing driver and constructor. Chinetti understood that his customer had to be special. Not just anyone could buy a Ferrari. The car would be defined in part by the man who sat in it.

Ferrari funneled every *lire* into the racing campaign. Money was tight, and the business model demanded that races be won. Why would a wealthy sportsman buy a Ferrari if a Jaguar had proved the finer machine on the track? It was survival of the fastest. Chinetti began funneling cash back to Italy, feeding the operation with a transatlantic umbilical cord. Nothing like a Ferrari had ever graced American roads. They were cars built by Italian artisans, every detail down to the steering wheel handcrafted using some of the same methods used to make Roman suits of armor and the royal carriages of the ancient kingdoms.

Three years after launching his company, Enzo Ferrari was poised to dominate racing in Europe. His first Grand Prix championships came back to back in 1952 and 1953. Another Le Mans victory came in 1954. Years later Ferrari was asked: Which of his cars was his favorite? He answered, "The car which I have not yet created." And which of his victories meant the most? "The one which I have not yet achieved."

On December 2, 1956, six months after the death of his son, Ferrari held the first meeting of his new Grand Prix team at the Modena

Autodrome. The mile-long track was carved out on the outskirts of town, roughly rectangular and flat, with a row of dilapidated pits that looked like an elongated bus shelter.

Hidden behind sunglasses, Ferrari greeted members of his new team—seven protagonists from all over Europe, each a national hero in his country.* Unlike Ferrari himself, his drivers personified the cars. They were young, aggressive, and impossibly handsome, all of them from money, for only the rich would have experience in performance cars. Here stood Alfonso Cabeza de Vaca y Leighton, the twenty-eight-year-old Marquis de Portago of Spain. Cigarette dangling, in need of a shave, he was audacity incarnate. "I like the feeling of fear," de Portago said in an interview that year. "After a while you become an addict and have to have it." In another interview: "Making love is the most important thing I do every day." Also mingling was Count Wolfgang Von Trips, nobleman of Germany. He was a Formula One rookie and would soon garner the nickname Count Von Crash.

A new season was about to begin. In Italy, the press baptized Ferrari's team *Il Squadra Primavera*—the Spring Team. The name would sadly prove ironic. One of the seven men would retire within days. All six others would be killed, all of them in high-speed crashes.

Meanwhile, at the Ferrari factory, Vittorio Jano was at work building a new racing engine, designed during those late-night meetings at Dino Ferrari's deathbed. Early dynamometer testing began before the year was out. In Ferrari's own words, a new six-cylinder engine the world would know as the Dino was about to "burst into song."

---

* According to Ferrari's friend and informal biographer Gino Rancati, who attended this meeting, the original seven were the Marquis de Portago (Spain), Count Wolfgang Von Trips (West Germany), Eugenio Castellotti (Italy), Luigi Musso (Italy), Cesare Perdisa (Italy), Peter Collins (Great Britain), and Mike Hawthorn (Great Britain).

3

TOTAL PERFORMANCE

FORD MOTOR COMPANY, 1963

> You go to a big football game. Say there are 100,000 people there. But not one of them wants to buy a goddamn football. You go to an automobile race and there they are — all your potential customers.
>
> — Ford racing executive JACQUE PASSINO

ON THE MORNING of February 24, 1963, beneath a blanket of Florida storm clouds, thousands of men, women, and children funneled through turnstiles and shuffled to their seats at the Daytona International Speedway. Detroit auto men turned up in droves. Henry II's brother Benson parted the hordes and took his seat trackside.

The 1963 Daytona 500 was America's first major speed competition following Henry II's withdrawal from Detroit's Safety Resolution. The publicity machine was in high gear. "A bitter controversy — beyond the point of intense sales competition — appears to be brewing in America's automobile industry," commented the *Los Angeles Times*. "Maybe today's race will touch off total war within the domestic car building ranks."

In the paddock, newspapermen could be seen peering under the hoods of a fleet of Ford Galaxie 500s, which were painted over with endorsements and racing numbers. There they witnessed for the first time the new Ford 427-cubic-inch V8, the largest, most powerful Ford production engine ever. No one at the time could imagine the impact this 427 would have in the world of speed and the future of the global car industry. General Motors—publicly an antiracing stalwart—had decided to stand by the Safety Resolution. Still, insiders believed the man behind the privately entered Chevrolets, Smokey Yunick, had his pockets lined with Detroit dollars. It was all a smokescreen. Yunick turned up with his own 427-cubic-inch Chevrolet engines. As the start neared, the sun peeked out from behind the clouds, warming Daytona's 2.5 miles of storied pavement.

This was American racing: an oval superspeedway and modified stock American cars, a perfect marriage between sport and industry. Swelled with cash, garnering larger and larger headlines, NASCAR basked in the Florida sunlight. Ford entered fourteen cars, and private teams entered fourteen General Motors cars. In the jostling for talent, Ford had signed a handful of first-rate oval racers—Fred Lorenzen, Ned Jarrett, and Dan Gurney. The Chevrolet and Pontiac lineups included A. J. Foyt, Fireball Roberts, and Junior Johnson.

By the time the pace car was leading the field of fifty gleaming stockers around the oval, 71,000 fans had begun to drain the speedway of its beer. In backyards and garages all over the United States, fathers and sons were tuning in by radio. The green flag gave way to an explosion of engines, and the fans followed the speeding metal around and around.

With fifteen laps to go, the crowd was treated to a breathtaking spectacle. Three Fords ran at the front in a tight pack, slipstreaming into the banked turns. Their chrome bumpers were inches apart, moving well over 160 mph. DeWayne "Tiny" Lund, a 270-pounder from South Carolina, was leading on the last lap when he ran out of gas in the final turn. His Ford coughed and wheezed and the engine fell silent just as he rolled over the finish line in first place, followed by four other Fords.

Down at Victory Lane, Lund clutched Miss Florida under his

arm and smiled for the cameras, the trophy sitting atop his Ford behind him, his wallet $23,350 thicker. Henry II's brother Benson headed for a pay phone.

It was Iacocca's job to turn success on the racetrack into success at the showrooms. If ever there was a time to do what he did best—sell—it was now.

Within days of the Daytona win, 2,800 newspapers featured advertisements that read: "In the open test that tears them apart—the Daytona 500—Ford's durability conquered the field: first, second, third, fourth and fifth." Ford's army of public relations men received a thirty-nine-page pamphlet on how to translate wins into sales. "Ford Motor Companies around the world have committed extensive resources to prove openly, dramatically, and conclusively on the international road and track circuit that its products are superior to those of its competitors," the internal document began. "The purpose is to help sell more Ford cars. Your job as a public relations man . . ."

Iacocca spruced up the line of showroom cars, adding glitz and velocity. The sober Falcon compact became the speedy Falcon Sprint. In the Galaxie, buyers could pay $461.60 extra for the 427-cubic-inch engine that debuted at Daytona. Iacocca toured the country holding press conferences about his "souped up, jazzed up" Fords, from a crowded Hollywood Palladium to Chicago to Detroit.

On Memorial Day, Iacocca took his seat at the Indianapolis Motor Speedway for the annual 500. Clutching a stopwatch, wearing a sport shirt with three buttons open at the top, he gazed down at the movable feast—America's most prestigious automobile race. Ford had partnered with Colin Chapman, the British founder of Lotus, to build two revolutionary rear-engined Indy cars. A reporter spotted Iacocca.

GM is attacking Ford over its investment in racing, he said. Any comment?

"If racing sells cars, what's wrong with that?" Iacocca answered. "It gives a guy who's going to shell out $3,000 a chance to measure the car's total performance."

He was careful to slip in those two words—"total performance"—Ford's new advertising motto.

As the green Lotus-Fords were wheeled onto the track, a hush fell over the crowd of nearly 250,000. A curious murmur followed. The Ford executive in charge of the Indy campaign, Jacque Passino, grabbed Iacocca's arm.

"Son of a gun, Lee! You hear *that*? They're all saying our name. That's what it's all about, man!"

Iacocca smiled. The excitement was tangible. "Forget all the details," he told Passino. "I'll take care of the details. Just go out there and race."

At the end of the day, Scottish driver Jimmy Clark took second in a Lotus-Ford behind Parnelli Jones's Watson Offenhauser in the fastest Indy 500 in history.

In Ford Motor Company's accounting department, a bespectacled numbers guy named George Merwin saw his desk towering high with paperwork. Henry II was spending unprecedented amounts on checkered flags those first twelve months, and it was Merwin's job to keep track of it all. Stock cars, Indy cars, travel expenses . . . Would the investment pay off? All the company had so far was publicity, which didn't add up in the accounting department.

Curiously, in the spring of 1963, the company began hearing extraordinary amounts of buzz from hard-core sports car enthusiasts. This struck everyone in Dearborn as odd. Ford didn't make a real sports car. Not yet, at least. That spring a car called the Cobra was crushing competition in Sports Car Club of America competition. It had a Ford engine in it, and it was built by a man named Carroll Shelby. Ford's investment in the car was negligible, and yet, Shelby's car was earning the company file cabinets' worth of news clippings.

Iacocca would never forget the day Shelby first walked into his office, a year earlier, in 1962. Ford's chief engineer Donald Frey made the introduction. Shelby was a tall Texan with a Lone Star accent—a "good lookin' son of a bitch," in Iacocca parlance. When they shook hands, Iacocca felt the calloused palm of a man who built cars, who had worked the Texas oil fields, and had competed in some of the most hard-fought automobile races throughout the 1950s. Shelby carried himself like a rich oil man, but as Iacocca would soon learn, he didn't have a dime.

The name Shelby didn't register with Iacocca or Frey. They were Detroit executives, not racing buffs. They didn't know Shelby was a champion driver, and that the thirty-seven-year-old had retired due to health reasons. That he was way down on his luck. As far as Iacocca was concerned, Shelby was just another would-be auto man showing up asking for money.

Shelby gave Iacocca the hard sell. By marrying a powerful American V8 engine to a small, lightweight European chassis, he could make a hell of a sports car for little money. "The idea is staring American car manufacturers in the face," Shelby said. "With $25,000, I can build two cars that'll blow off the Corvettes."

Blow off the Corvettes? Iacocca thought to himself. *The undisputed king of American sports cars?*

By the time Shelby was done with his pitch Iacocca was sold. It wasn't just the idea he bought into. It was the man. Shelby was instantly likeable. He spoke at a high volume from years of trying to talk over revving engines, and he conveyed in his voice a sense of what was at stake in it all for himself—something far more than money. It took one genius of a salesman to recognize another, and there they were, Lee Iacocca and Carroll Shelby staring at each other.

"Give him the money and get him outta here," Iacocca told Frey, "before he bites somebody."

Ninety days later, Shelby was back in Dearborn with the first Cobra. The chassis came from A. C. Cars of England, a small financially ailing company that made roadsters and wheelchairs for invalids. The engine was a new small-block Ford V8, a cheaply made but wonderful lightweight high-performance power plant. The Cobra had a low profile, with gills on its side, glittery wire-spoked wheels, and a long nose that suggested plenty of muscle under the hood. It appeared to be moving even while it stood still.

"You know," Frey said, looking at the thing, "I think you might be onto something here."

Shelby's bills were added to the pile on George Merwin's desk in accounting.

Shelby came from a desiccated East Texas dot-on-the-map called Leesburg, population two hundred. He was the son of a horse-and-buggy mail carrier. As a child he spent hours sitting by his door watching the first automobile traffic motor by on the newly surfaced road in front of his house. His father took him to his first automobile race. Shelby's father told him, "There's no man born with a drop of red blood in his veins that doesn't enjoy a race of some kind."

World War II gave Shelby his first taste of speed. He joined the Air Corps and was stationed at Lackland Army Air Force Base outside San Antonio, where he learned to fly B-25s, B-26s, and later B-29s. When the war ended he got married, had three kids, and set out looking for a way to make money. He started a dump truck business but it failed. He worked the oil fields. He tried his hand at raising chickens but ended up with twenty thousand dead birds, killed by a case of botulism. One day a friend named Ed Wilkins asked Shelby if he'd drive a car he'd built in a competition. That's when it all began.

"Drive it?" Shelby said. "Where?"

"There's a drag meet in a few days at the Grand Prairie Naval Station, between Dallas and Fort Worth."

"You mean just straight dragging on a strip?"

"Yeah. It's a quarter-mile and I'd sure like to see how this baby goes."

"You just got yourself a driver."

Shelby made the fastest time that day. Winning suited him, and he found himself in the right place at the right time. As expensive European sports cars appeared in the United States for the first time after World War II, a road racing Renaissance took root. Rather than oval speedways, these drivers were racing on twisty road circuits like in Europe. At twenty-nine, Shelby began showing up with sports cars owned by his friends at races all over Texas and Oklahoma. One 100-plus-degree August day in 1953, Shelby was working on his farm and found himself late for a race at Eagle Mountain Naval Station in Fort Worth. He arrived just in time and competed while still wearing his striped farmer overalls. He won the race, but the overalls stole the show. They became his trademark.

A rich patron named Tony Parravano took Shelby overseas for his first time, in 1955. Parravano, a thirty-something-year-old bowling-ball-shaped man, had made millions developing land in Southern California. Rumor had it that he was connected to the Mob, and in two years' time he was to disappear forever, his body never found.

"I'm going to buy another 15 Ferraris," Parravano told Shelby before they left for Italy, "and I'd like you to drive them for me. Any of them, I don't care. Take your pick."

Shelby spent a month hanging around the Ferrari factory while Parravano negotiated the deal. When *Il Commendatore* learned that the Texan was one of the hottest talents in America, he requested a meeting. Ferrari spoke in Italian and his secretary translated. He offered Shelby a contract—50 percent of purses plus a few *lire* a month.

"You can drive some sports car races," Ferrari said.

"With all due respect, Mr. Ferrari," Shelby responded, "I got three kids at home. I make more money driving in the States."

Ferrari was incredulous. No one ever turned down a spot on his team. "What's this about money?" he said. "You're at the beginning of your career. You should be honored to drive for us."

Shelby faced Ferrari down, then stood up and left.

Two years later, Shelby rolled the dice overseas with the Aston Martin team. In Europe, he became part of the international fraternity of elite racing drivers. He was so affable, even his competition adored him, and he could drive like a comet. His face appeared on the cover of *Sports Illustrated* in 1957: Sports Car Driver of the Year.

One morning in 1960, Shelby awoke with a sharp pain that felt "like a knife being stuck in my chest." He was diagnosed with angina pectoralis, a heart ailment in which the coronary arteries are starved for blood. For the next few months he lived on a steady diet of nitroglycerin tablets, tucking pills under his tongue during races to keep the engine in his rib cage from overheating. He drove his last race on December 3, 1960, placing fifth in a Maserati at the *Los Angeles Times-Mirror* Grand Prix. At thirty-seven, he found himself out of work. He never did make much money—$75,000 his best year. Good money, but not enough to retire.

Shelby had three kids, an ex-wife, and dead chickens in the bank. One night he was in Lake Tahoe as a guest of William Harrah, the casino magnate and a real car freak. Shelby met a beautiful brunette named Joan. They sat together drinking. Care for another? She was gorgeous and he "could talk the leaves off the trees," as one acquaintance put it. He had this idea, he told her, an idea he'd been tossing around for some time.

"To build my own sports car . . ."

With a tiny investment from Ford Motor Company, Shelby hired his first employee, his new girlfriend Joan, and got himself an answering service. He built the first Cobra out of the back of Dean Moon's hot-rod shop in Santa Fe Springs. Iacocca came through with more money, and Shelby moved into a shop at 1042 Princeton Drive in Venice, California, not far from the little house he lived in near the beach in Playa del Rey.

Ford Motor Company displayed the $5,995 Cobra—with its "Powered by Ford" logo just aft of the front wheel—at the 1962

New York Auto Show. With Ford money Shelby purchased his first advertisement. It was tucked into the October 1962 issue of *Playboy,* amid news of "The Return of the Ascot" and a new "tart-tongued" *Tonight Show* host named Johnny Carson. On the racetrack, the Cobra performed the impossible, beating the Corvette Stingrays. The Dearborn suits couldn't believe their eyes when *Road & Track*'s June 1963 issue landed on their desks: "It seems that in a ridiculously short time, the Corvette has been clouted from its position of absolute primacy."

Chevrolet's Corvette had been a thumbtack on every Ford executive's seat since its debut ten years earlier.

When Iacocca flew out to have a look at Shelby's operation in Venice, he found an amazing grease-stained facility on a sun-baked back street. There was an open-door policy. You could walk right in. "Hi, I'm Lee," Iacocca said, introducing himself to Shelby's employees. They knew who he was. They liked that he didn't have any pretensions. Iacocca knew better than to wear a suit. He showed up in shorts.

Shelby took Iacocca around. Racing cars were everywhere, in various forms of undress. Piles of Goodyear tires reached up to the ceiling. There was a dyno room, inside which engines shrieked day and night. In the metal shop, workers were fabricating parts by hand right there.

Most of Shelby's employees looked as if they were straight out of high school. They weren't college types, but they were street smart. Shelby called them "hot-rodders trying to prove that they weren't the dipshits everyone in the world thought they were." They knew how to weld, how to fabricate, how to make cars go fast. Walking around the shop, hearing the hiss of air hoses and smelling the sweat and oil, Iacocca knew that Shelby had tapped into something very real and powerful.

"I'm impressed," Iacocca said.

"Shucks," Shelby said, "I'm not an engineer. I'm not even very smart. The only thing I understand is human nature. I just like to bring the right people together and see what happens. I think I've put the right people together at Shelby American."

In the summer of 1963, one year after Ford Motor Company backed out of the Detroit Safety Resolution, racing drivers turned the nation's tracks into action-packed advertisements for Ford cars. Dearborn began to hear buzz from the dealerships. Cars were rolling out of showrooms at a speedy pace. As the economy started to look up, Americans went on a car-buying binge, setting a blistering sales pace that threatened to beat the record year of 1955. Sunday after Sunday, Ford and "Powered-by-Ford" Shelby Cobras claimed more trophies, and each Monday, Iacocca followed the numbers on the upswing.

As the summer ended, the *New York Times* made it official with a story about Ford's soaring sales. It appeared on the front page. "Does winning automobile races sell cars?" the article began. "You bet it does." The piece called the success of racing as a marketing tool "immediate and remarkable."

Outside Henry II's corner office window, the stacks of the Rouge spewed smoke at full bore. But Henry was not in his office. Many executives noticed he had been spending far less time in the Glass House. In fact, the Deuce had set his sights on something far more ambitious than anything on Lee Iacocca's radar.

On a drizzly June afternoon, Henry II landed in a chartered jet in England and arrived at the Ford factory at Dagenham outside London. When a reporter asked about his visit, he answered, "I came to Europe to see what was becoming of our investments, which between 1960 and 1964 will have totaled $800 million."

Followed by an entourage of company board members, Henry II disappeared inside Dagenham's fiery belly. What the Rouge was to America, Dagenham was to Europe—an industrial metropolis, the largest factory on the continent. Its belching stacks towered over the River Thames about eight miles east of London's city center. Henry II could remember standing on this piece of earth back in 1928 and surveying the horizon, seeing nothing but desolate marshlands. He was eleven years old when he watched his father Edsel break ground at Dagenham with a ceremonial shovel, inaugurating "the new Detroit of Europe."

Now, thirty-four years later, the plant had been expanded and modernized. Its 170 gigantic presses turned 600 tons of sheet metal into car bodies every day. Roughly 1,200 Ford Cortinas were rolling off the assembly line every 24 hours, headed for the artificial rainstorm facility, where they were tested for leaks before being shipped to dealerships all over Europe.

Twelve hundred Cortinas a day. The numbers were staggering, but Henry II knew it wasn't going to be enough. Projections pointed at an incredible upsurge in demand for automobiles in Europe. Unlike the United States, Europe had staged most of World War II and its economies and infrastructures had been devastated. Finally the continent had caught up. What occurred in America in the 1950s was about to happen in Europe in the 1960s—a crystallization of the middle class and all that came with it. A booming economy was going to fuel a car-buying binge among a huge population, many of whom had never owned an automobile before.

That spring, Henry II was overseeing the most ambitious expansion project in his family company's history, and almost all of it was in Europe. A plant in Halewood outside Liverpool would be operational by October with eleven thousand workers. A new foundry was in the works near Dagenham with a power station strong enough to supply electricity to a city of 160,000 (such as Modena, Italy).

Henry II envisioned a world in flux. His industry no longer had borders. For the first time, foreign cars posed a major threat to his market share in America, from expensive Jaguars to cheap Volkswagens. (He is rumored to have called the Volkswagen "a little shitbox.") And in Europe, American car companies were battling for customers like never before. Both Chrysler Corporation and General Motors were investing wildly in the market overseas. Henry II upped the ante, spending hundreds of millions to buy back shares of European Ford subsidiaries in Denmark, the Netherlands, Belgium, Italy, Spain, France, Britain, and West Germany. He was consolidating power.

Even then, as he toured Dagenham, he had in his mind an idea: to launch the first ever pan-European automobile company, a highly coordinated company that would produce and sell affordable cars in noncommunist nations from England to the border of Russia. He knew the day would soon come when Ford Motor Company stood to profit as much in Europe as it did in America. Maybe more. He needed to send a message.

Ford cars were the best in the world.

Since his earliest days, Henry II's father Edsel had opened his eyes to what the continent had to offer. Europe was a place of beauty, romance, fantasy. Through all his suffering until his death, Edsel Ford had found in Europe his escape. He adored the old world's sense of aesthetics and its luxurious automobiles. The home he built for his family in Grosse Pointe was a monument to Europe. The roof's stone tiles, stained glass windows, and much of the furniture came from Europe. The main staircase where Henry II first learned to climb steps had been shipped across the Atlantic.

Like his father, Henry II found himself seduced by Europe. Only in his case, it wasn't cars or masterworks of art that hooked him. It was a woman. Far from Dearborn, where his every move and every shift of the stock price were scrutinized, Henry II was living a secret life.

Her name was Cristina Vettore Austin. She was a thirty-six-year-old Italian divorcée living in Milan. Henry II had met her at Maxim's in Paris. With his wife and children, he had been invited to a party in honor of Princess Grace of Monaco. He spotted Cristina: an exquisite woman with green eyes, earthy dark blond hair, and a model's figure. She was mingling that night among Europe's inner circle of wealthy jet setters. Henry II switched the place cards at the dinner table so his was next to hers. She spoke English with a charming Italian accent.

Soon Henry II was seeing his Italian "Bambina" frequently—one night in Paris, another in Geneva. He set up an apartment for her at 530 Park Avenue on the southwest corner of 61st Street in Manhattan, across from his suite at the Regency Hotel, on the northwest corner. They sailed the Mediterranean on his new yacht, christened *Santa Maria*. (One person who saw Cristina said that she "looked better in a bikini than anyone I've ever seen.")

The more time Henry II spent with her, the more he feared the consequences: his marriage of twenty-one years and three children, and his reputation. Not only was he having an affair, she was a *foreigner*. There were rules of society. In many regards, the year 1963 was still a part of the 1950s. You might've slept with another man's wife, but you didn't tell anybody. You didn't mingle in the wrong circles and you didn't wear a tie that was more than two inches wide. Divorce was against the rules, and breaking the rules was risky for a man who ran the world's second-largest company.

Henry II began to explore the line that separated himself from his public life. In Dearborn he was Mr. Ford. In Europe he could be Henry. For the first time since he had taken over Ford Motor Company in 1945, the interests of these two men were not the same.

"I've got the company," he said one night while discussing marriage with his mistress. "There's you and Anne, but I'm really married to the company. That's the one relationship in my life that will always be there."

During the summer of 1963, Mr. Ford's employees stepped carefully around him. He seemed increasingly erratic. Bottles of scotch emptied quickly in his presence. One day during a meeting in Dearborn, Henry II sat listening to Iacocca pitch the idea of a new sporty model in front of the executive committee. This was Iacocca's baby, the Mustang project.

"Don't give me this shit, Lee," Henry II said. "Just don't talk to me about it." He paused. Then he said, "I'm leaving." He stood up, walked out, and headed to the hospital. Doctors claimed he was ill with mononucleosis, but a rumor spread that Mr. Ford was on the verge of a nervous breakdown.

For Henry II, success in Europe was critical. Therein lay the legacy that he would leave behind. Meanwhile, blind items had begun to appear in the gossip columns about a married American industrialist and his secret lover. Business and personal life, Europe and America — it became apparent to Henry II that his two worlds were going to collide.

"He was like a time bomb," one Ford executive later recalled. "You could almost hear the ticking."

FERRARI, DINO, AND PHIL HILL

1957–1961

> I was just a young man caught in a dangerous occupation and needing to go on. Should you want confirmation of what it's like, ask a combat veteran from Vietnam or another conflict how they managed to go on in the face of death. And if you think there's a great difference between them and an impassioned race driver because the latter has the ability to quit at any time, you don't fully understand the pull of being impassioned.
>
> —Phil Hill, 2004

Following Dino's death in June 1956, a string of tragedies began to plague Enzo Ferrari's factory in Maranello. Some called it "a jinx." Others were less kind in their assessment of the facts.

The troubles began on May 12, 1957, at the Mille Miglia, a race on 1,000 miles of public roads in Italy, with Brescia at the top (the start and finish) and Rome at the bottom. Some ten million spectators lined the course, forming a barrier on each side of the pavement. Before the 1957 Mille Miglia, Ferrari summed up his favorite event to a reporter: "It is the race of the people. One may say that the whole of Italy leans forward with her eyes on the tarred strip of road somewhere along the course on Mille Miglia day. It is a day when I feel my life is useful."

Before dawn, under klieg lights in Brescia, thousands turned up to watch sports cars launch from a starting ramp. A Ferrari won the race as usual, but the victory was eclipsed by news of a tragedy, news that quickly spread from Tokyo to New York. Racing driver Alfonso de Portago, Marquis of Spain, was moving at near top speed in a Ferrari 335 through the rural village of Guidizzolo, when he blew a tire. The Ferrari guillotined a telephone pole, swerved headlong into a crowd, and came to rest in a drainage ditch.

Twelve were killed and many more were injured. De Portago's fame made the story all the more shocking. He was a nobleman and an international celebrity, an Olympic bobsledder, the married lover of both actress Linda Christian and Revlon model and fashion icon Dorian Leigh. He was a character so romantic and mysterious, even Ferrari found him fascinating. (Ferrari described him as "a magnificent brute.") At the scene of the accident, de Portago was found twice. His body had been severed in two.

Ferrari routinely read seven or eight newspapers a day, and he recoiled from the venom in the press. "The last time." "Abolish the Mille Miglia." "Enough with these absurd races of suicide and massacre." The Mille Miglia—for thirty years Italy's most beloved sporting event—was never held again. The Magician of Maranello bore the brunt of the blame. He was served papers: "Enzo Ferrari, born in Modena on the 20th February, 1898, and resident therein, [is] charged with manslaughter and causing grievous bodily harm by negligence . . ." The charge claimed that Ferrari had used Englebert tires not equipped to handle the speed that his cars could travel. He would be found innocent, but the charges would hang over him for the next seven years.

When hearings began, Ferrari defended himself. He was a fierce nationalist, and he had dedicated his life to winning races in honor of Italy. He felt betrayed by his country.

"Why should I continue in an activity whose only reward is being branded a murderer?" he asked.

He threatened to quit, but he couldn't walk away. Before the 1958 season, he debuted a new racer—the Dino Formula One car. In its nose was the six-cylinder engine designed on Dino Ferrari's deathbed. Like Dino himself, the car was destined for death and glory.

July 6, 1958: Luigi Musso of Rome was killed in a Dino on the treacherous Muizon corner at Reims during the French Grand Prix. August 3: British driver Peter Collins followed, also in a Dino, at the German Grand Prix at the Nürburgring. At the Grand Prix of Morocco, the final race of 1958, Ferrari contract driver Mike Hawthorn placed second, earning enough points in the Dino to secure the World Championship. It was as if Ferrari had resurrected his son and made him a champion machine. Weeks later Hawthorn's body was found lifeless in a wrecked Jaguar beside a wet highway outside London.

Two of the original seven Spring Team members remained.

A debate over racing's death toll raged in Europe among sportsmen, journalists, moralists, and politicians. One estimate put the figure at 25 percent each year; one in four Grand Prix drivers who started a season could expect to be dead before it was out. After each tragedy came the grim cleanup, and the photos in the papers of weeping mothers and beautiful widows in black dresses.

How to regard athletes willing to give their lives in the pursuit of glory?

The racing driver spoke of speed's sensuality, the way a lifetime of experience can be piled into a single lap around a circuit. And the challenge to one's psyche—to master the car was to master oneself. There was the story—most believed true—of the driver pulled from a wreck with two broken legs, ankle, nose, ribs, three vertebrae. Doctors checked him over: his pulse and blood pressure were normal. When one had gone so deep into oneself, was there a way back out? To paraphrase the writer Ken Purdy: quitting was in itself a form of suicide. "Only those who do not move do not die," the French driver Jean Behra said after de Portago's death. "But are they not already dead?" (Behra crashed a Porsche at the Grand Prix of Berlin on August 1, 1959, and was hurled from his mount. A witness described his final moment: "For an instant he could be seen against the sky with his arms outstretched like a man trying to fly.")

And the cars: Was the will to achieve progress, ever-increasing speeds, undermining itself?

Every year engineers summoned more power from engines while lightening the cars. The results were automobiles so fast, the slightest error or unforeseen variable could be catastrophic, and the idea of safety—seat belts even—was considered unmanly.

"The modern racing car has become like the old Miura bull," claimed retired Ferrari driver Piero Taruffi in an article in the *Saturday Evening Post* entitled "Stop Us Before We Kill Again!" "It has developed into an animal that is too high powered. The racing car must be bred-down to a safer speed." These vehicles required "almost a superhuman quality from the race car pilots," according to the *Corriere della Sera*.

Though other teams suffered casualties, none suffered such loss of life as did Enzo Ferrari's. The Vatican publicly attacked Ferrari in its official newspaper *l'Osservatore Romano,* calling him "a modernized Saturn turned big industrialist [who] continues to devour his sons. As it is in myth, so too is it in reality."

Ferrari publicly declared his bottomless grief for each of his drivers killed. He would remember shaking their hands for the last time. At the factory his inner circle saw him struggle to maintain composure, pulling himself together to maintain his public persona. But his critics argued otherwise. They said Ferrari controlled his drivers like marionettes, the strings being their own talent and vanity. They said he created rivalries between his men to spur them to move faster.

"I wouldn't want to be in the shoes of those boys," said veteran driver Harry Schell in 1959. "When you drive for Ferrari you are headed one way only: for that little box under the ground." (Schell was killed in a Cooper at Silverstone a year after uttering these words.)

No one could argue that Ferrari's cars were less safe than any other. To the contrary, neither he nor his technicians ever cut corners. When asked about the root of his mania, his obsession with victory, Ferrari told one reporter in 1958, "Everything that I've done, probably, I did because I couldn't do anything less . . . One day I want to build a car that's faster than all of them, and then I want to die."

"It does not seem to me that I have ever committed a bad act," he would later write in his memoirs. "Yes, I am calm, even if not serene, even terribly imperfect, as I am. I have never repented. I have regretted often, but repented never. Is this a good thing? I fear not. I feel myself to be alone after so many delirious events, and almost guilty for having survived."

On the eve of the 1960s, Ferrari found himself lacking talent. He had one young pilot on his team who'd been clamoring for a shot at the big time. The racer had been hanging around Modena for a couple of years, competing in smaller events and working as a test driver. He was a rare breed in this part of the world—an American. With each death, the American climbed a spot on the team. Suddenly, he found himself at the top.

"How would you like to drive for me at Le Mans?"

And there it was, a dream and a nightmare embodied in a sentence. Enzo Ferrari had spoken it in French. The man sitting before him—Phil Hill of Santa Monica, California—didn't speak Italian, but he spoke French. They were sitting in Ferrari's office at the factory on either side of an expansive desk. Hill was smaller than the old man; he stood five feet ten, his body cut wiry and strong, with short arms and heavily muscled hands and wrists. He was twenty-eight, but his short-cropped hair, smooth skin, and shy brown eyes made him look younger.

Hill answered yes—he would like to drive for the team at Le Mans. But even then, Ferrari sensed hesitancy. The equivocation was clearly articulated.

"What do you think of our great protagonist?" asked Ferrari. He was speaking of Umberto Maglioli, an Italian who had beaten Hill at the Mexican road race weeks earlier. Hill had come in second.

"Oh, I think he's just fine," Hill said, trying to sound enthusiastic.

"Well, I'm going to team you both in the same car at Le Mans. What do you think of that?"

"Oh. Oh that would be great. That would be fine. I look forward to it."

The invitation was not just to drive at Le Mans. It was to join the team, and it had been accepted.

"Well," Ferrari said, "come now and let me show you the great cars you will be driving."

Phil Hill followed his new boss into the Ferrari racing shop. A spot on Ferrari's team was uncharted territory for an American. Hill was the first.

Hill was born on April 20, 1927, into a wealthy family. As a boy, growing up in Santa Monica he didn't have a good relationship with his parents, and he was awkward socially. He was self-conscious, withdrawn, and he wasn't good at sports. Other kids laughed at him when he tried to swing a baseball bat, and the snickering scarred him. Instead of playing with other children, Hill liked to wander alone in junkyards studying the rusted heaps. He could identify any make of car at a glance by age six.

Hill's aunt bought him his first car when he was twelve—an old Model T for $40—and with the help of his aunt's butler, he overhauled the engine. He was too young to drive on public roads, but he had a friend named George Hearst, grandson of the publisher William Randolph Hearst, who had a horse track on his family estate in Santa Monica Canyon. The boys spent hours on the dirt oval, sliding into turns and speeding the straights.

Hill dropped out of the University of Southern California and got a job selling foreign cars at Roger Barlow's International Motors. He began to refine his crude racing skills. He competed at Torrey Pines, Palm Springs, Santa Ana. The road races drew large crowds and drivers from as far off as Tucson and San Francisco. Far from meccas like Le Mans and Monza, a group of talented young men was creating a European-style sports car racing scene in the dusty flats of Southern California. As Carroll Shelby became the man to beat in the South and Midwest, Hill won race after race on the West Coast. He read everything he could about his heroes—the ace pilots of Europe.

Both of Hill's parents died in 1951, and with his inheritance in hand, he contacted Luigi Chinetti in New York, the Ferrari distributor. Chinetti offered Hill a bargain: $6,500 for a 212 Export. The used Ferrari cost three times what a new Ford would, and it had a checkered past. That spring, a driver named LaRivière had crashed into a wire fence at Le Mans and had been decapitated. The dents had been hammered out; Enzo Ferrari needed the money. And so Hill began showing up at races in his Italian roadster, the third or fourth Ferrari ever to roll onto California asphalt by his estimate.

In two years' time, Hill was competing in international events. Racing side by side with his idols, the Californian found that he could compete. But these races opened his eyes to a different world. At a race in Argentina he saw a driver crawl out of a wrecked Aston Martin in a ball of flames. The man burned to death right before Hill's eyes. Hill competed at Le Mans for the first time in 1953 in an Osca. Days later, he attended the funeral of American racing driver Tom Cole.

The racetrack was a place for ruthless men, like Alberto Ascari, who was mean to his own son so the boy wouldn't miss his father when he died in a racing accident, which he did in a Ferrari at Monza. And the Frenchman Jean Behra, who wore a prosthetic ear due to a crash, and liked to remove it when the comic effect was opportune.

The stress took its toll. Hill suffered anxiety attacks before races. When he smoked his last prerace cigarette, his heart would be fluttering, muscles twitching. His stomach grew so raw he was forced to eat baby food. Even when he won, he suffered fits of depression.

What was he doing risking his life? He didn't fit the stereotype of the wild-man American racing driver. He was from a socially prominent family. He spoke French and adored Beethoven. And yet he was compelled to compete.

Hill had X-rays done on his stomach, and his doctor warned him about an ulcer. So he quit racing at twenty-six. Months went by.

One fall day, with the 1954 Carrera Panamericana approaching—the Mexican road race—Hill received a letter from a Texas oil man named Allen Guiberson, who contracted jockeys to race his expensive mounts. Inside the envelope was a photo of a 4.5-liter Ferrari with a shark tail fin on the rear deck. Guiberson had scribbled five words:

"Guaranteed not to cause ulcers."

The Mexican road race ran across the Pan-American Highway, from the Guatemalan border to the Rio Grande, broken up into stages. Five days, 1,908 miles. "You go as fast as you think you can," Indy driver Chuck Stevenson said, describing the ride. "And then you go even faster." Hill placed second in the race, which claimed seven lives. The Mexicans had a nickname for him: *El Batallador.* That's when he got the call from Mr. Ferrari. Soon he was on board a swaying train headed south down the Italian Riviera, en route to Modena.

Contract signed, Hill moved into the Hotel Reale on the Via Emilia. He began to spend hours at the Autodrome testing Ferraris, and keeping fit by riding a bike in the hills west of the factory. At night the Reale's bar was the place to be. The saloon crowded with drivers, women, wealthy car buyers, journalists, crackpots, and hangers-on, all lured to the Italian city by the bellow of the 12-cylinder engine.

Hill was "on the end of the bench," in his own words, suffering an inferiority complex. The men he mingled with were teammates, but he knew they'd do what they had to do to defeat him on the track. He raced at Le Mans, the Nürburgring, Monza. He traveled with a piece of hi-fi equipment and his favorite recordings of Beethoven and Vivaldi.

Ferrari's engineers learned to respect the cagey American's talent. He was a skilled technician, and he began to pick up the Italian language, which endeared him to the mechanics. He understood the limit of his car and the limit of his abilities, intimately enough to reach both while staying in control.

In some ways, Hill fit in here more so than at home. Racing drivers were respected professionals in Europe. It was a vocation, one for educated men who wore ties to dinner. Only, unlike opera singers and professors, these men were trailed by death constantly. The danger didn't seem to bother other drivers, but it bothered Hill. He had evolved, as one profiler put it, "into one of consummate and meticulous skill nurtured by an increasingly heightened and almost paranoid fear of death."

"I would so love to get out of this unbent," he told one reporter. "I have a horror of cripples. Even when I was a little boy I couldn't bear to look at anyone who was deformed, could not bear to see them suffering. I guess I've always worried about ending up that way myself."

Hill always had to keep his eyes open, to make sure he didn't become the victim of another man's mistake, another driver or a mechanic. He didn't have much contact with Ferrari. When he found himself in Ferrari's office, he saw the black-framed portraits on his desk—de Portago, Musso, Hawthorn. They were champions and martyrs. Hill felt a strange tension between himself and his boss. Perhaps they didn't entirely trust each other. America was Ferrari's biggest market, and Hill possessed intrinsic value. He was opening the eyes of an increasing population to the glamour of European motor sport, and to the beauty of Ferrari cars.

The U.S. fans waited to see if Hill would fulfill what they believed was America's destiny: a World Champion of their own. Or would he end up one more black-framed portrait on Ferrari's desk?

In 1958, Hill opened the season with wins at the 12 Hours of Sebring and the 1,000 Kilometers of Buenos Aires. In the spring, Ferrari teamed him with the Belgian war hero Olivier Gendebien at Le Mans. The pair would share a 250 Testa Rossa ("red head," as the 3-liter V12's cam covers were painted red). No other race carried the commercial significance that this 24-hour event did. Ferrari's entire business plan was based on proving on the racetrack that his cars were special—not just beautiful, but the finest and fastest in the world. Victory at Le Mans' *Grand Prix d'Endurance* was proof.

When Hill climbed from his cold sheets on race day, he saw rain clouds outside the window. Standing in the pit before the start, he turned to his teammate.

"We can win this race if we have the guts to go slow the first part," he said to Gendebien.

The Belgian took the initial shift. Hill stood in the pit watching and smoking, awaiting his turn at the wheel. He took over and, at nightfall, those rain clouds opened up. With almost zero visibility, Hill unleashed one of the greatest performances in Le Mans history, passing car after car in the pounding rain. At dawn, broken cars lined the course, one engulfed in flames. A dozen drivers had crashed, one fatally. The #14 Ferrari was well in the lead, and hour after hour, Hill came closer to immortality.

On June 22, 1958, at 4:00 P.M., Phil Hill became the first American Le Mans champion. As his Ferrari spirited him past the checkered flag, he threw his fist in the air in triumph. He drove the victory lap with his teammate sitting on the hood of the car. The rain had cleared and legions of fans watched the exhausted pilots cruise by, knowing that, with an American champion crowned for the first time at Europe's greatest race, American cars couldn't be far behind.

THE PALACE REVOLT

ITALY, 1961

> This kind of love, which I can describe in almost a sensual or sexual way within my subconsciousness, is probably the main reason why, for so many years, I no longer went to see my cars race. To think about them, to see them born and see them die—because in a race they are always dying, even if they win. It is unbearable.
>
> —Enzo Ferrari

On a winter day early in 1961, journalists gathered in the courtyard of Enzo Ferrari's factory to see the unveiling of a new fleet of racing cars. The group shuffled around on the cobblestones, sizing up the machinery. In due time, Ferrari appeared. His press conferences were odd affairs. The man remained aloof, though he understood the power of the typewriter as well as anyone. These new cars, he said, married his time-honored philosophy—the importance of pure horsepower—with the most cutting-edge innovations.

"I want to create a car," Ferrari said, "with the greatest possible speed, the least weight, the least fuel intake, and all parts of equal durability."

The last three years had seen an unprecedented leap forward in racing car technology. The machine that resembled its prewar ancestors morphed into a Space Age bullet. The concept of aerodynamics was new. Designers had begun to change the shape of cars so that they could cut through air efficiently. Disc brakes, developed in England, allowed drivers to wait until the last split second before slowing into a turn. England's revolutionary Cooper Formula One cars had proven the efficacy of mounting the engine *behind* the cockpit. A rear engine allowed designers to shave off pounds by simplifying the driveline, lower the center of gravity to within inches of the ground, and keep the weight of the power plant directly over the driving wheels. All of this served to place more performance in the hands of the driver.

Ferrari was responsible for none of this innovation. But married to the awesome power of his engines, this new technology would result in unprecedented dominance.

Phil Hill realized early in 1961 that the Ferraris were not going to lose. He braced himself. There was going to be an intense struggle for victory within the team. He could not know at the time that the season would take an unexpected turn, that the foundation of Ferrari's empire was about to be cracked.

In sports car racing, Hill won at Sebring in March. In June, he won at Le Mans. Ferrari's cars placed first, second, and third. The nearest competitor, a Maserati, finished 183.92 miles behind. Ferrari had claimed the Le Mans crown three of the last four years. Hill had the perfect temperament for endurance racing—a fierce competitiveness joined with great mechanical sympathy.

In Formula One, however, the drama proved more complicated. The new Dino F1 car was blood red and rear engined. Its front end had dual air intakes, resembling the gaping jaws of a feeding shark. Thus its nickname: Sharknose. The Dino was going to chariot one driver to World Champion honors, but the team featured two senior members—Hill and the Count Wolfgang Von Trips. They were equally matched.

It would be *mano e mano,* just like Ferrari liked it.

Count "Von Crash," West Germany's most celebrated sportsman, was an affable thirty-three-year-old whose good looks and ability to speak four languages made him an ambassador for his sport. He possessed none of Hill's mechanical intellect, but he had the gift of natural instinct and an absolute adoration for the glamour of speed racing. In May, Von Trips won the Dutch Grand Prix in the Dino. On the podium he stood next to Hill, who'd placed second. Mobbed by fans and photographers, the blue-eyed Count pulled Hill in and laid the victory wreath over the two of them, a gesture of solidarity. At the Belgian Grand Prix a month later, Hill won, followed by Von Trips. At the British Grand Prix, it was Von Trips, then Hill.

The two men were friends. As the season progressed, however, their teeth clenched behind their smiles. No American had ever won the Formula One title. No German ever had either. Behind the scenes, the tension at the factory was excruciating. Hill was a master of speed. Von Trips was willing to give everything.

That summer, mainstream America opened its eyes to the European racing scene for the first time. Hill was regarded as the homegrown hero, a world-class racer but also a deep-thinking man, an intellectual. His thirty-four-year-old face, wrinkled now, with creases of thick flesh around his worried eyes, showed up in countless newspapers and magazines. The *Los Angeles Times* called the kid who couldn't swing a baseball bat "Mickey Mantle in a Ferrari." *Esquire:* "He is resolve, terror and courage all in one." *Newsweek* put Hill on its cover. The story quoted Enzo Ferrari: "Of course I am concerned about my drivers. Each time I shake hands with one of them and give him a car, I wonder: Will I ever see him again?"

The season came down to one race, the Italian Grand Prix at Monza. A victory by either man would clinch the title. During practice the day before, Enzo Ferrari stood in the pit. Each day before the Italian Grand Prix he made an appearance. It was one of his only public appearances of the year and it caused a stir. Ferrari wouldn't stay for the race; he'd be leaving soon. In the pit, Hill's nerves were stretched to the breaking point. He was talking to his mechanics about his windscreen. It was poorly designed, he argued. The Dino was not perfect. Ferrari's angry bark interrupted.

"Maybe you ought to just put your foot down harder!"

On Sunday afternoon, September 10, 1961, Hill stood behind the pits with his helmet on while an aid poured water from a can down the neck of his coveralls. The cool water helped to calm him. He preferred to start the race soaking wet. Thousands had crowded Monza's woody acreage awaiting the race's start. It was a great moment for Italy, a chance to see a Ferrari pilot clinch the F1 World Championship in front of a home crowd. Around the circuit television cameras were mounted; the race was one of the first to be broadcast.

Wearing a silver helmet in honor of West Germany, Von Trips settled into his Dino. Monza was not his favorite track. He had crashed here before, nearly been killed. Twice. In qualifying, he'd won the pole. Hill and the rest of the drivers settled into their cars. Engines running, earplugs in, goggles on, hands on the throbbing wheels. In front of his television, Enzo Ferrari took his place.

The starter waved the Italian flag and the pack of red, green, and silver fuselages fired forward, trailing a cloud of exhaust. Gaining velocity, they headed for the first turn, the *Curva Grande,* then around turn two into a straight. Hill made his move early. From out of the second row, he darted into the lead. Von Trips fell behind, battling with a Scottish youngblood named Jimmy Clark in a lime-green Lotus. They hurtled around the circuit, Hill in front, Von Trips and Clark dueling nose to tail a few lengths back, over the dangerous Monza banking,* past the pits and into lap two. On the backstretch, the pack screamed down the straight headed for the tightest turn—the 180-degree *Parabolica,* a sweeping second-gear curve with a blistering approach.

* The Monza Autodrome had a steep American-style banked turn. The 1961 Italian Grand Prix was the last time it was used, as it was deemed too dangerous. It still exists at the track, covered in weeds and hidden in the woods. Looking at it today, it's hard to fathom that men actually raced cars on it.

Hill downshifted and entered; Von Trips and Clark followed. At well over 100 mph, Von Trips went for the racing line. So did Clark. Their wheels touched. The cameras were rolling. Von Trips's Ferrari spun out of control and hit the inside guardrail. It cartwheeled up a five-foot grass embankment and, twisting like a helicopter blade, the red Dino scythed through the ranks of spectators standing behind a wire fence before tumbling back toward the pavement.

What had happened? Smoke obscured the chaos in the crowd. A flag man came sprinting, signaling caution. The race went on.

Forty-one laps later, Hill crossed the finish line first and the checkered flag waved. He knew there'd been an accident, and he had a good idea of who it was. But he never removed his eyes from the action unfolding directly in front of him. He pulled into the Ferrari pit and saw technical director Carlo Chiti.

"And Trips?" Hill asked. "Is he dead?"

Chiti's face was grave. "Come on," he said, "they want you for the awards ceremony."

The next morning, in a hotel in Milan not far from the Autodrome, a new World Champion descended a flight of stairs into the lobby. Hill found a crowd gathered around the hotel's only television. They were watching Von Trips's accident, which was being aired over and over. Footage of the ace pilot being carried off on a stretcher, his arm dangling off the side, drew gasps. A commentator groaned on in Italian. Count Wolfgang Von Trips and fourteen spectators had been killed.

Hill found a place to sit and a friend spotted him. It was Robert Daley of the *New York Times*. Daley had been on the European racing beat for awhile and had befriended Hill.

"Are you going to quit, Phil?" Daley asked.

"I don't know," he said. "I haven't made up my mind yet." After a pause, he became philosophical. "After all," he said, "everybody dies. Isn't it a fine thing that Von Trips died doing something he loved, without any suffering, without any warning? I think Trips would rather be dead than not race, don't you?"

"What are you going to do, Phil?"

Hill thought for a moment. Then he said, "When I love motor racing less, my own life will be worth more to me, and I will be less willing to risk it."

In Maranello, Enzo Ferrari stayed secluded behind his factory gates. He too had seen the accident on television, and when the news arrived—the death toll—he prepared himself for the fallout. The front page of the *Corriere della Sera:* "Fifteen Dead from the Tragedy at Monza. The Investigation Has Begun at the Autodrome."

In the past, accidents had always been mysteries, with variables never fully understood. They unfold so quickly and violently, no two witnesses tell the same story. "Unfortunately," Juan Manuel Fangio once said, "the man who could tell you exactly what happened is always dead." But this was different; television cameras had filmed the last race of Count Wolfgang Von Trips. A government commission of experts promised to place blame. "You can't imagine what it was like," Hill later recalled. "It seemed like everybody in the damn country was milling around Maranello, and there's Enzo Ferrari, with three days' beard growth, and wearing bathrobes all day."

Ferrari did not attend Von Trips's funeral, though his wife did. Hill was a pallbearer. The casket was driven through the streets of the Count's home city of Cologne in the pouring rain atop his overheating Ferrari roadster. The procession of umbrella-holding mourners snaked as far as the eye could see.

The most dominant year in Ferrari's history ended in despair. Every victory and every tragedy since Dino's death had heightened the tension inside the factory. And in the fall of 1961, the situation in Maranello deteriorated. Dissatisfied employees began to make demands. Ferrari's wife, they said, was nosing around too much, and there were other complaints. Perhaps the bad publicity and the bloodshed was too much for them to handle. Rumblings inside the factory grew in volume.

Nothing like this had ever happened, as Ferrari's word was always the last. Now major figures in the organization were bonding against him. As a result, eight key men left him in November, two months after the Monza tragedy. Among them was Ferrari's racing team manager and chief engineer. The company's finance guru, its foundry manager—all left at once. Ferrari refused to beg anyone to return. He called a meeting of his junior staff.

"We got rid of the generals," he said. "Now you corporals must take charge."

Ferrari's company was struggling. He spent all his profits on racing, and he was badly in need of money. He was being vilified at every turn. In the newspapers that he read religiously, the Magician of Maranello was being called the Monster of Maranello. One contract driver's wife called him "an assassin." Had he not brought Italy another World Championship? Had he not raised the reputation of the Italian automobile into the stratosphere? He believed his country had forsaken him, and his own men had betrayed him, walked out in a much-publicized "Palace Revolt."

The old man was livid. And so he came up with a plan.

FERRARI/FORD AND FORD/FERRARI

SPRING 1963

The American really loves nothing but his automobile.

—William Faulkner

In February 1963, officials at Ford's German division in Cologne received a mysterious letter from the German consul in Milan. The letter told of a "small, but nevertheless internationally known Italian automobile factory" that was for sale. No names were mentioned. The letter fell into the hands of Ford's Director of Finance in Cologne, Robert G. Layton, who at first believed the Italian factory must be one of the foundering operations sideswiped by an unstable economy. Some legwork, however, revealed that the factory in question was producing the most famous sports and Grand Prix cars in the world.

Layton forwarded the note to Dearborn on February 20. "For what it is worth, I am attaching a letter," his communication read. "While I doubt whether this is of special interest, there may be angles that I do not know of."

By coincidence, Lee Iacocca and his chief engineer Don Frey had discussed the possibility of purchasing Ferrari. Most Americans had never seen a Ferrari in person, but they knew of its status. A mention of those three syllables signified beauty, sex, money, fame. Most of all: *speed*. The only way Iacocca could bring that kind of clout to Ford was to buy it.

When he approached his boss with the idea, it struck a chord.

The notion sat well with Henry II's plans for Europe. He knew intimately the mystique of this Italian automobile. In 1952, Enzo Ferrari gave the grandson of Henry Ford a black 212 *barchetta* as a gift. It was a beautiful car of clean, simple lines. Henry II added his own touch, a little hint of the Stars and Stripes. Around the Italian wire-spoked wheels, he fit a quartet of Firestone whitewall racing tires.

If the market abroad was the future, and racing victories translated into sales, the Ferrari factory could be a brilliant strategic acquisition. Iacocca got the go-ahead to explore.

On April 10, 1963, the phone rang in the office of Enzo Ferrari's right-hand man, Franco Gozzi, at the factory. A thirty-one-year-old six-footer known for his quick wit, Gozzi was a Modenese trained in law. He was married to the daughter of Ferrari's barber Antonio.

Gozzi struggled for a second. He had a coffee in one hand and a crossword puzzle in the other—a rare moment of relaxation, as the boss was not around. When he answered the phone, a voice came through in Italian with an American accent.

"Filmer Paradise here, is Mr. Ferrari in the office?"

Gozzi replied that he would take a message.

"Is Ferrari there or not?"

"No."

"Tell him I called to know whether or not we can arrange a meeting."

When Ferrari returned to the factory and got the message, he thought quietly for a moment. Something was up. He knew Filmer Paradise—president of Ford Italiana. "Let him come," Ferrari told Gozzi. "Fix a meeting in the old office in Modena, maximum secrecy, inside the company too."

Ferrari and Paradise met two days later. Yes, Ferrari said, he was interested in striking a deal. He chose Ford because he was a great admirer of Henry I, he said. He would sell his factory to the Americans as long as he could retain control of the racing team. He had no interest in the customer cars.

"I never felt myself to be an industrialist, but a constructor," Ferrari told the Ford man. "The production development of my firm is only of interest to me if conducted by others." Then: "But be quite clear that in the construction and management of the racing cars I want absolute autonomy."

Paradise reported back to Dearborn. The next day, a group of fact finders prepared to head to Italy. Among them was Roy Lunn, a thirty-eight-year-old Englishman with a taste for fine suits and pocket squares. Lunn was the one Ford engineer in America who had experience building racy European automobiles. When he was in his twenties he had worked at Aston Martin, where he designed the DB2, which won Le Mans in its class in 1950 and 1951. Lunn's little Aston became the "it car" to own in Europe as a result. Lunn joined Ford in 1953. By some accounts, he had the best job in Detroit. He was paid to draw up plans for vehicles of the future. Among his design concepts was a three-wheel flying car topped by a helicopter propeller and a 170,000-pound supertruck called Big Red. He was Ford's answer to Q in Ian Fleming's 007 novels, complete with patrician British accent.

If the deal with Ferrari came to fruition, Henry II was likely to install Lunn as chief engineer in Italy. Before he left for Europe, Lunn met with Iacocca in his office and Iacocca warned him of a hostile reception.

"It would be like an Italian Mafia group came to the U.S.," Iacocca joked, "to buy the New York Yankees."

On April 13, Lunn checked into the Hotel Reale on the Via Emilia in Modena.

Lunn and four other Ford men arrived at the factory gate in two cars. Gozzi was there to meet them, and soon Ferrari made his entrance. He had a way of stripping people of their confidence with a mere handshake. A prominent Italian journalist once described Ferrari: "The Drake was big and robust and with his way of doing things seemed like a monument, before whom everyone, absolutely everyone (heads of state, kings, businessmen, actors, singers and all others) felt great embarrassment."

Ferrari led the Ford men into the works. In the racing department, a large open bay with a two-story-high ceiling, sports cars and Grand Prix machines were situated in a diagonal row, tended to by men in gray jumpsuits. Windows allowed natural light to illuminate the workspace. The facility was stark and small. On the other side of the factory was another open bay—the assembly line, where the top-of-the-line 400 Superamerica was in production. (*Car and Driver*'s take on the $17,800 Superamerica that year: "Owning one is, or should be, the goal of every automotive enthusiast anywhere.") Machinists worked over presses, lathes, and drills fashioning cylinder blocks and crankshafts. Each man was in charge of a piece of equipment, with directions to work according to a specific drawing on the machine with absolutely no deviance. The little factory was the antithesis of the American plant. Everything was done by hand, with no automation. That year, Ferrari's four hundred–plus employees constructed some six hundred customer cars. The Rouge produced twice that many cars every day.

When Ferrari and Gozzi showed the group Maranello's wind tunnel, the Americans chuckled. It was fashioned from a defective Ferrari engine, a rotor, and a large, white, horn-shaped funnel. Unlike the advanced technology Ford had access to in America, Ferrari's wind tunnel resembled an oversized tuba, and it was big enough to test the airflow over diminutive models only.

When Lunn saw the foundry, he noticed that, like the rest of the place, it was immaculate. The whole factory seemed the only place in Italy not dusted with tobacco ash.

"Boy is it clean," Lunn remarked. "It's amazing that you can keep the foundry like this."

Ferrari answered through an interpreter: "I have taught these men how to wipe their asses."

In early May, Iacocca's number two, Don Frey, arrived from Dearborn to close the deal. The asking price was $18 million, a pittance considering Henry II's personal fortune, estimated at nearly half a billion dollars.

A thirty-nine-year-old with dark hair parted neatly on the side, Frey had the face of a man ten years his junior. Shortly after their meeting, however, Ferrari learned that behind Frey's fresh face and dainty glasses lay an exacting and brilliant mind. Frey had earned a PhD in metallurgical engineering from the University of Michigan. His love of cars was as renowned at Ford as Iacocca's ability to sell them.

Frey and Ferrari spent hours together talking, engaged in long discussions about the difference between American and Italian machinery. Ferrari doodled logos, trying to join the two company names. He called the Ford executive *Dottore Ingegnere,* a sign of respect for the American's intellect. They took drives in Ferrari cars up into the Apennines. On these roads, with their loops and switchbacks, Ferrari made use of his old racing skills. "He drove like a mad man," Frey later remembered. "He loved to get me in the car and try to scare me."

As negotiations moved through May, Frey reported back to Henry II. Never did it occur to the Ford men that the whole deal could in fact be an elaborate machination, a ruse. That Enzo Ferrari may have had another agenda completely. By this time, news of the deal was public knowledge, making headlines all over the world. It would be the most unusual merger in automotive history. The Italian press was up in arms. It was as if they were losing a national treasure—the Ferrari automobile—to these arrogant Westerners.

This pleased Ferrari. It seemed he was not the Monster of Maranello after all, but a monument to Italy. No Italian would ever underestimate his value again.

On May 21, both parties sat down with a final contract (one version in English, another in Italian) in Ferrari's office at the factory. On one side of the table sat Don Frey and a sizeable delegation of Ford lawyers. On the other was Ferrari, one lawyer, and an interpreter. From his portrait on the wall, Dino Ferrari looked on. The final asking price was down to $10 million—a sum so small, it didn't make sense. And yet, no one at Ford questioned it. The agreement called for two entities. *Ford-Ferrari:* A company, 90 percent of which would be owned by the Americans, which would manufacture customer cars with a "Ford-Ferrari" badge on them. *Ferrari-Ford:* The racing team, 90 percent of which would be owned by Ferrari.

Dearborn's brainpower had concluded that the number one priority of this latter organization would be Le Mans. This 24-hour test in France carried more commercial importance than all other European races combined. At Le Mans, Ferrari was invincible. Buying Ferrari meant buying Le Mans.

As Ferrari read the final document, the Ford men saw him underlining certain passages in violet ink. In the margin, he drew a large exclamation point. It was clear that he was angry.

"But here," Ferrari said to Frey, holding the document. "It is written that if I want to spend more for racing I have to request authorization to do so from America! Is it also written that way in the official English text? Where is the freedom that I demanded right from the start to make programs, select men and decide on money?"

"But Mr. Ferrari," Frey said, "you're selling your firm, and you pretend still to dispose of it to your pleasure."

Ferrari paused. "*Dottore Ingegnere,*" he began, "if I wish to enter cars at Indianapolis and you do not wish me to enter cars at Indianapolis, do we go or do we not go?"

Frey replied without hesitation. "You do not go."

Ferrari's face contorted. Something uncoiled inside him. "My rights, my integrity, my very being as a manufacturer, as an entrepreneur, cannot work under the enormous machine, the suffocating bureaucracy of the Ford Motor Company!" he shouted. This was followed by a lengthy invective. Gozzi, who was present, described it as "a tirade that I had never seen or heard before in my entire life and have not done so since."

When Ferrari was through, he turned to Gozzi. It was 10:00 P.M. "Let's go and eat," he said. The two men rose and left the room. The negotiations were over.

The next day, Frey departed Italy with a signed copy of Ferrari's book, *The Enzo Ferrari Memoirs: My Terrible Joys,* which had just been published, as a parting gift. A long flight later, he found himself sitting in Henry Ford II's office on the twelfth floor of the Glass House. Henry II's blank stare asked the question: Well?

"Mr. Ford, I failed," Frey said.

The two men went upstairs to the penthouse dining room, where Frey gave a blow-by-blow description of his Italian sojourn. He'd never been in the penthouse dining room. He would later describe that meeting as "the longest lunch I ever had in my life."

Henry II listened. He did not like what he was hearing. No Italian hothead was going to get in the way of his plans.

"All right," Henry II said. "We'll beat his ass. We're going to race him."

The statement lingered in the air. Slowly it came to Frey, a vision of what was about to happen and what his role in it would be.

"How much money do you want to spend?" Frey asked his boss.

"I didn't say anything about money," Henry II answered.

Frey got the point: Whatever it cost, Henry II was going to foot the bill. There'd be no excuse for second place.

When Iacocca got word, he immediately had a memo sent to his staff: "Prepare a presentation of plans we propose to implement in view of the suspension of Ferrari negotiations." A week later he had a proposal in hand to set up a new special-vehicles department unlike any other in Detroit. Its purpose: *to design and build the fastest, most reliable and technologically advanced racing car in history.*

America was going to go to Europe and beat Enzo Ferrari at Le Mans. In a Ford.

PART II

A CAR IS BORN

MEANS AND MOTIVE

> To take control of this materialized energy, to draw the reins over this monster with its steel muscles and fiery heart—there is something in the idea which appeals to an almost universal sense, the love of power.
>
> —*Motor World*, 1901

THE LAST AND ONLY time an American car won a major European race was in 1921, forty-two years earlier. With half his body wrapped in bandages following a crash in practice, San Francisco-born Jimmy Murphy won the French Grand Prix in a Duesenberg. Drivers had attempted to win the 24 Hours of Le Mans in American cars before, in Chryslers, Overlands, even a Ford in 1937. In the early 1950s, Briggs Swift Cunningham ("Mr. C." to his friends) built a series of eponymous cars that represented America's most heralded challenge at Le Mans, finishing third in 1953 and 1954. The best finish of any American car was a Stutz in 1928—second place.

Henry II had resolved to do what had never been done. At stake was the reputation of his family legacy, Ford cars. Not since Henry Ford unleashed his flathead V8 engine—his last great invention—in 1932 had the family empire enjoyed a reputation for innovative engineering. Henry I's grandson was going to change the reputation of the company forever—if in fact he could win.

The Deuce soon learned that he had added incentive to beat Ferrari. Vengeance became part of the mix. Not long after the Ford-

Ferrari deal went south, reports of Ferrari-Fiat negotiations became public. For years, Fiat had provided Ferrari with a stipend, a cash bonus for his work bringing prestige to the Italian automobile. Now, in 1963, Ferrari was beginning to lay a foundation for a deal with Fiat that would supply him with a windfall of badly needed *lire.* Was Ferrari working a backdoor deal the whole time? It became obvious to many that Enzo Ferrari—a nationalist to the core—never had any intention of selling to a foreign entity. Certainly not to Americans, who had in fact bombed the hell out of his factory during the Big One.

Ferrari had toyed with Henry II while the world looked on. The stage was set for a war of speed.

They say automobile racing is as old as the second car. Since Karl Benz first patented the "motorwagen" in 1886, cars evolved into two diverse species on either side of the Atlantic. In America, with its vast roads, mapped out by urban planners who literally moved mountains to make way for them, cars were all about the big engine. Racing stock American cars on an oval track was a tradition that reached back to 1896. With grandstand seating, a promoter could funnel spectators through turnstiles and charge them to see the show. Spectators could witness the entire race rather than a small slice of it. Oval track racing was staged theater, a performance before a paying audience. Whether it was comedy or tragedy depended on the day.

Europe, in contrast, was the cradle of racing. Town-to-town road races spread the gospel of the automobile across the continent. In contrast to America, roads in Europe molded to the contours of the earth, with twists, bends, hills, dales. Cars evolved with smaller engines capable of quick bursts of power. Constant acceleration/deceleration and cornering required durable gearboxes, supple suspension, quick steering mechanisms, and long-lasting brakes. European automobiles evolved into lighter, more sophisticated devices. Twisty racing circuits were the laboratories where new technologies were tried and proven.

In 1922, two Frenchmen came up with an ambitious idea: to hold the ultimate motor race. Charles Faroux was a brilliant engineer and France's doyen of motoring journalists. Georges Durand ran France's Automobile Club de l'Ouest. Their idea was to arrange a 24-hour contest that would test every facet of an automobile. The endurance race would reveal a car's weaknesses. Racing at night by headlamp would force competitors to improve upon primitive electrical systems. The winning car would be not only the fastest, but the most fuel efficient and durable, overall the most intelligently engineered.

Faroux and Durand chose to host their race outside Le Mans. It was here that men came to risk their necks while testing the boundaries of the physical universe. France held the first Grand Prix ("great prize") here in 1906. Two years later, Wilbur Wright took to the skies above Le Mans in the first significant European flight. Faroux and Durand charted a course on public roads outside of town, which came to be known as le Circuit de la Sarthe, for the nearby Sarthe River. A Chenard et Walcker won the 24 Hours of Le Mans in 1923, averaging 57.2 mph.

W. O. Bentley was the first to understand the commercial importance of this race, the prestige that came with victory, not just for the make of car but the country it came from. His English touring cars achieved a dynasty at Le Mans, winning from 1927 to 1930. Alfa Romeo followed—1931 to 1934. The French cars of Ettore Bugatti dominated the prewar years. Following World War II, Enzo Ferrari emerged as the greatest of all Le Mans constructors. In each case—Bentley, Alfa Romeo, Bugatti, Ferrari—Le Mans victories were responsible for the status of these budding companies and created demand for these cars across Europe. If it were not for these checkered flags, would these companies exist today?

Ferrari called Le Mans The Race of Truth. Over 24 hours, two men (one in the car at a time, the other filling up on nutrients in a catering tent, or catching a bit of sleep in a trailer) traveled the 8.36-mile circuit attempting to go faster than all others. The car that completed the most laps at the end won. By the early 1960s, competition had been divided into two basic classes: Grand Touring cars or GTs (production cars that customers could purchase) and Prototypes. While both were required to have headlights, a two-seat cockpit, and trunk space, the Prototypes were in fact purpose-built racers. Prototypes topped GTs by some 70 mph in flat-out speed in some cases. They were meant to exemplify the sports cars of the future, the science of speed pushed to its limit in street-legal cars.

In Dearborn, on July 12, 1963, Lee Iacocca held a meeting of his executive committee. Don Frey opened with a presentation. The plan was to launch a highly specialized division that could focus on building a prototype Le Mans car. Roy Lunn would oversee engineering. The new division, called Ford Advanced Vehicles, would take advantage of the company's financial resources while operating with the agility and freedom of a small, independent manufacturer.

"The objective," Frey said, "is to have a car running in one year."

In other words, in time for the 1964 Le Mans.

Lunn took the floor. He reported on a trip to Europe days earlier. At the 1963 Le Mans, he watched Ferrari cars place first, second, third, fourth, fifth, and sixth. He shared his analysis with his colleagues: To beat Ferrari, their racing car would have to travel at top speeds over 200 mph, with the durability to race for 24 hours. (At that speed, a car covered the length of a football field in a single second.)

"With the exception of land-speed record cars," Lunn said, "no vehicle has ever been developed to travel at speeds in excess of 200 mph on normal highways. These speeds are greater than the take-off speed of most aircraft, but, conversely, the main problem will be to keep the vehicle on the ground."

To build such a car, all the most advanced design technology and aggressive innovation would come into play. The engine would be located amidships, behind the cockpit and ahead of the rear axle, just like the Ferrari 250 P, the winning car in 1963. Every weld would have to hold. At Le Mans, no weakness could hide.

"Attempting to meet these objectives represents a scintillating technical challenge," Lunn continued. "The competition has reached its sophisticated product level after nearly 40 years of evolutionary development."

Lunn proposed a seven-figure budget and was amused at how quickly Iacocca approved not just this figure, but the entire plan. All agreed this new racer would have to be built in Europe initially, as few of the specialized components needed were available in the United States.

Though his name was not mentioned during this meeting, there was one man everyone present knew would have to be brought into the fold. No one at Ford had ever built a sports car that could beat the Chevrolet Corvette at the track, let alone a Ferrari. There was only one man in America who had the knowledge and experience to build and develop a winner. He wasn't exactly the corporate type.

By the summer of 1963, Carroll Shelby's company was on sound footing, its tiny assembly line churning out hand-built "Powered by Ford" Cobras. Shelby's list of customers included Wilt Chamberlain, Bill Cosby, Vic Damone, James Garner, and the "King of Cool" Steve McQueen. The Cobra had captured the imagination of a new breed of speed enthusiast. There was even a Top 40 hit blasting over the radio waves about Shelby's car, "Hey Little Cobra," recorded by the Rip Chords.

Like Ferrari, Shelby had created a business model that depended on winning races. If the cars didn't win, the cars wouldn't sell. His company lived and died on the track. That summer, in Sports Car Club of America competition, the Cobras were winning *everything*. Shelby had all the reporters in his pocket. Los Angeles alone had four dailies, the *Times* and the *Daily News* in the morning, and the *Mirror* and the *Herald-Examiner* in the afternoon. The words "Carroll Shelby" were thumping out of those journos' typewriters one after the other.

Around the corner from his facility at 1042 Princeton Drive, Shelby had added another shop on Carter Street. His men were no-names in the world of international racing, and yet, they were a font of unearthed talent. Chief engineer was Phil Remington, a forty-two-year-old who was working at the Princeton Drive shop for the previous owner and kind of came with the place when Shelby rented it. Remington was a mastermind with no college degree or any formal training, who'd been building speed machines since the early days of hot-rodding on dry lake beds northeast of Los Angeles. Competition manager was Ken Miles, a World War II tank commander from England. At forty-five, Miles was four years older than Shelby, a sharp-tongued engineer who'd earned a reputation as the best 1.5-liter-class racer in California and perhaps anywhere.

Recently, Shelby had hired his own photographer so he could supply shots to the papers and magazines. Dave Friedman's darkroom at the Carter Street shop was actually an old toilet and smelled like one. Shelby loved to look over Friedman's shoulder watching blank sheets of paper sit in the pools of chemicals. Slowly an image came into focus: cars being fabricated, cars at speed.

Shelby knew he had to take the Cobra overseas, where the Ferrari was king of the road. There was opportunity in Europe for a renegade American car manufacturer. Shelby knew as well as anyone the significance of victory at Le Mans. For men who built cars, it was the biggest stage in the world. The last time anyone beat Ferrari, the year was 1959 and Shelby was at the wheel of an Aston Martin. He became the second American after Phil Hill to win the 24 Hours of Le Mans. During the entire race he wore a pair of chicken-farmer overalls.

Throughout Shelby's years with Aston Martin in the late 1950s, he had developed a distaste for Enzo Ferrari. More than once Ferrari had offered Shelby a contract. And each time the Texan had turned it down. Perhaps he had a sense that signing that paper would be like making a pact with the devil. Yes, Ferrari's cars were the greatest. But Shelby saw the men on Ferrari's team die one after another in 1957–1958. He believed *Il Commendatore* was responsible.

"That son of a bitch killed my friend Musso," Shelby said about the Roman pilot Luigi Musso, who died in a Dino at the French Grand Prix. "And he killed others too."

In the spring of 1963, Shelby organized a press conference. When he called a meeting, everyone showed up. He stood before reporters and photographers, stared right into the eye of a television camera, and announced that he was forming a new team to take on Europe. The Texan was gunning for the Monster of Maranello. There was something between Shelby and Enzo Ferrari, Shelby's girlfriend Joan later said. "It went way back into Shelby's early racing career, and it was very personal." And so the first major factory-backed American campaign to win the 24 Hours of Le Mans would be two-pronged. Shelby had filed paperwork to homologate his Cobra to race in the GT class against the Ferrari GTOs, while Ford would build a prototype to try to win Le Mans outright.

"Next year," Shelby announced, "Ferrari's ass is mine."

IL GRANDE JOHN

ITALY, SPRING 1963–1964

> The highlight of my career was the moment that sparked it all, at 17, winning a relatively unimportant road race in Wales riding a Vincent which I largely built myself. It was the first time I felt man and machine came together. I'll never forget the vibe I felt knowing I was part of the bike. It flowed . . . I was flying . . . That was the day the future was born.
>
> —John Surtees

On a winter morning early in 1964, Enzo Ferrari emerged from his new home at 11 Largo Garibaldi in Modena. A purring Fiat 1100 was waiting, with a loyal chauffeur named Pepino at the wheel. At sixty-six, after so many years of success and tribulation, Ferrari was finally enjoying a little personal wealth. His new home—around the corner from the Orsis' manse, owners of Maserati—was immense. There was more than enough room for Ferrari and his wife, and his mother, Mama Adalgisa.

Ferrari stepped into the car. He preferred to be driven now, though the Fiat remained his transportation of choice. "What's so surprising about that?" Ferrari said. "You're just as likely to see [Fiat scion] Giovanni Agnelli driving a Ferrari." He had been given an honorary doctorate in mechanical engineering by the University of Bologna and now preferred to be addressed as *Ingegnere* rather than *Commendatore*. Or simply, Ferrari. Otherwise, his routine hadn't changed. Pepino delivered his boss to Antonio D'Elia's barbershop, where gossip was exchanged. *Quei Canarini non possono segnare uno scopo!* Then it was off to Dino's tomb. After that he headed to the factory.

In 1964, the world of speed and sports cars had gone mainstream and global, and competition for customers was extraordinary. Porsche had just debuted the inimitable 911, and Jaguar's E-type, the most phallic car ever to roll down a road, was a hit. That spring Ferruccio Lamborghini unveiled his first customer car: the $13,900 Lamborghini 350 GT. It was an exceptional debut with a high-revving 12-cylinder engine. Still, no car inspired the kind of awe that Ferrari's two new 1964 models did.

The 275 GTB was deemed "the Sophia Loren of supercars," impossibly elegant and yet built for speed. The new 500 Superfast was the top-of-the-line Ferrari—a $29,300 customer car that could top 170 mph. Only thirty-six Superfasts were made. *The New York Times:* "Just to sit in one feels a bit dangerous." Peter Sellers purchased a 500 Superfast. The Shah of Iran bought two. As Roberto Rossellini put it: "There is no finer thrill in the world than driving a Ferrari flat out."

Ferrari had become arguably the most famous man in Italy after the Pope. Indeed, he was sometimes referred to as The Pope of the North, and his Vatican City was Modena, "a noisy nirvana of automobiles," as one writer called the place. That he was a visionary and a genius was indisputable, and he knew it. Some 40 percent of his cars were purchased in North America, where clients now included the Dupont and Dulles families, New York's Governor Nelson Rockefeller, James Stewart, William Holden, Steve McQueen,* and Hollywood film director John Frankenheimer.

Phil Hill had faded from the Italian scene. Following the death of his teammate Von Trips, Hill faced disappointment at every turn. During the 1962 season, the team's new manager Eugenio Dragoni complained over and over to the boss. Ferrari received

* McQueen's 1963 Ferrari 250 GT Berlinetta Lusso sold at auction in 2007 for $2.3 million.

phone calls from Dragoni in postrace reports from racetracks all over the continent.

"Sì, sì . . . sì, sì. Ma il tuo grande campione non ha fatto niente." *Your great champion didn't do a thing.*

Hill had little choice. He left the team. Even as a champion, he had never enjoyed Enzo Ferrari's favor. "I wasn't sorry to leave," Hill later said in an interview with a reporter. "Enzo Ferrari never understood me . . . He always favored the man who would take that extra risk in a live-or-die situation. I wasn't willing to die for Enzo Ferrari. I wasn't willing to become one of his sacrifices."

Italy had embraced the new pilot at Ferrari. From the beginning, his presence shook things up in a way no other Ferrari contract driver ever had. A thirty-year-old Englishman, his name was John Surtees. The fans called him *Il Grande John*. He had thinning dark blond hair, pale English skin, and a body so thin his clothes draped over him as if on a hanger. Tucked behind a striking nose, he had movie-star blue eyes. Surtees would never forget the day he met Enzo Ferrari. "It was a curious feeling as I walked through the door of Ferrari's office that morning, as if I was stepping into another world," he later recalled. "It seemed as though everybody was going about their jobs with a reverential earnestness which was almost unnatural. I was experiencing for the first time the unique magnetism of Ferrari."

The first meeting took place at the cramped old *scuderia* office. Ferrari eyed the young driver from his chair behind his desk. He knew he was staring at a question mark. Surtees was a motorcycling World Champion. Only recently had he added two more wheels to his repertoire. He had but one full season of races under his belt in cars. But Ferrari saw something in him, a bottled-up fury that reminded many of "The Flying Mantuan" Tazio Nuvolari. Surtees was pure aggression. Ferrari bypassed the pleasantries.

"I would like you to drive for us next year," he told Surtees. "Formula 1, sports cars, and anything else we might decide to race. Here's the contract."

Surtees faced one of the most difficult decisions of his life. He turned down the contract so he could gain more experience. The offer came again a year later. Ferrari was at the time rebuilding his team, with new engineering talent in the form of a local Modenese named Mauro Forghieri and a new team manager from Milan named Dragoni. With Hill gone, Ferrari needed a driver to anchor the team and Surtees signed on.

Surtees was born in 1934, the son of a talented motorcycle racer. He spent much of his youth in his father's motorcycle shop in a rural village south of London. During World War II, Surtees's father trained dispatch riders for the British army. These men would carry messages by motorcycle from the officers to the soldiers on the front lines. It was deadly work, and the faster and more skilled these riders were, the better chance they had at staying alive. The war robbed Surtees Sr. of his racing prime, something his son would never forget.

At fifteen, Surtees left school and took on an apprenticeship at the Vincent motorcycle factory in Stevenage. The entire family—mother, father, and three kids—led a nomadic existence on weekends, caravanning from one race to the next all over Great Britain. Money was always tight. John and his father raced Nortons and Vincents. Mother's official job titles were "chief mechanic" and "caterer."

When he was twenty-two, Surtees was hired by Count Domenico Agusta to race MV Agusta motorcycles, which was based outside Milan. By age twenty-six, Surtees had won seven Grand Prix World Championships for Count Agusta (four in the 500-cc class and three in the 350-cc class). Though an Englishman, he'd become a national hero in Italy. The Italians grew accustomed to seeing the young champion of speed in their sporting pages, his body coated neck to toe in skintight black leather, straddling his machine. During the season, he lived for much of the time out of hotels in Italy, and he began to pick up the language by watching spaghetti westerns at the cinema. He learned to eat Italian, to think Italian. He was blazing a path. For centuries, the British had come to Italy to be enchanted. Surtees came to become the fastest man on two wheels in the world.

He had never *seen* a racing car up close the day he stepped in one and motored onto the track at Goodwood in England. Six months later, he appeared at his first Formula One race, driving a Lotus. In his third Grand Prix he won the pole in Portugal. The press made no bones: Surtees's ability to compete on four wheels instantly was unprecedented, miraculous. "He doesn't seem to know the meaning of fear," said manager of the Vanwall team David Yorke, who gave Surtees a tryout. "And yet at the same time he isn't reckless either."

In spirit, Surtees belonged in Italy. When he joined the Ferrari team before the 1963 season, some didn't take to him. He was inexperienced on four wheels, and yet, he was so fast he bruised many an ego. Unlike the prototypical Ferrari driver, he didn't come from money. He was a gritty working-class bloke with icy blue eyes and a blazing temper, the first of a new breed of racing's angry young men.

Surtees debuted with the Ferrari factory team at the 1963 12 Hours of Sebring, winning the race. "Boy, if Horatio Alger could only have seen John Surtees hurling a Ferrari down the road," commented *Car and Driver.* His contract didn't pay him much. He had a retainer and earned a percentage of prize money. It added up to a decent wage. But in Italy, with its anemic *lire,* the living was good.

Ferrari sold Surtees a gray 330 GT at a discount, which had enough room in it for his racing kit and luggage. He roomed at the Hotel Reale, where the bar was still haunted by Mike Hawthorn, who before his death held court nightly there, drinking cherry brandy and smoking Senior Service cigarettes. A night at the hotel cost two thousand *lire,* barely more than one English pound. At the Tucano, Surtees could fill a table with all the classic Modenese dishes for the equivalent of fifty pence. He traveled the world with his wife Pat, who kept his lap charts, from Monte Carlo to Florida to South Africa.

From the beginning, and unlike Phil Hill, Surtees had an excellent rapport with Ferrari. The two were seen lunching frequently in town. Racing gossip could chew up the afternoon hours. The Englishman's spot on the team was an experiment that Ferrari set in motion. Surtees was trying to become the first man ever to win World Championships on two wheels and four. He was game, and the fans were riveted.

On the tail of his first season with Ferrari, Surtees went west on an expedition to America, in the fall of 1963. He competed in a new Ferrari Le Mans-type sports car called the "P car" (for prototype) at Mosport in Canada and Riverside in California. He returned to Italy with insight into the American scene. There was a blossoming world of sports cars in the United States. Some sharp young engineers were building experimental lightweight cars, European in style, but with the American way of thinking: the big engine. Guys like Carroll Shelby and his Cobra, and a Texan named Jim Hall and his Chaparral.

"We cannot compete with the big engines being used in America," Surtees warned Ferrari. Surtees also feared that the team was not evolving fast enough. There was a time, in the very recent past, when Italy was the center of innovation. Ferrari, Maserati, and Lancia: the legendary marques competed against each other, but also copied each other's best ideas. Now, with Ferrari's exception, those factory-backed racing teams had ceased to exist, due to the wobbly *lire* and the bloodshed. The best ideas were now coming from the fiercely competitive young teams of England. And who knew what kind of threat those big American engines might pose someday?

In Italy, there was only Ferrari, and the old man had his old way of doing things. Surtees could see a time in the future when the competition would overwhelm Ferrari like a gigantic wave of fast-moving metal.

"Look," Surtees told Ferrari upon his return from America, "we are in a desert here." He paused. *"We are all alone."*

They had to innovate, think young and fast. They had to become something new, or else.

One day early in the spring of 1964, Ferrari's chauffeur drove him to the Modena Autodrome to observe a shakedown of the new prototype Le Mans car. As the Fiat pulled into the parking lot, Ferrari saw the small crowd gathered in the pit area. He stepped out of the Fiat and heard the wail of the racing car's V12 engine. From the pit Ferrari watched Surtees navigate the circuit.

The 1964 Ferrari 330 P was an update of the prototype sports car that'd conquered Le Mans the year before. The engine was mounted behind the cockpit. The four-liter V12 with six twin-choke Weber carburetors delivered roughly 370 horsepower in dyno tests, extreme power for a car that weighed 1,665 pounds with a full tank. At the wheel, Surtees's spine soaked up the engine's throb. To create this machine, Ferrari had wrung every drop of imagination from some of the greatest mechanical minds in Italy. He once described the process of building such a car:

> In the first act of his labor, the maker conceives what his creature is to be: he dreams of it and sees it in detail, and he lays down the plan of work which he entrusts to a band of helpers who share his passion. A racing car, in fact, does not necessarily come into being as the creation of a superior mind, but is always the compendium of the common unflagging and enthralling work of a team of men fired by a common enthusiasm. There follows its construction, which must nearly always be done in record time, although it never takes less than six to eight months of feverish work. The next stage is the assembly of the car and its testing, which is the most delicate, the most engrossing and the most dramatic phase . . .

This dramatic phase had begun. It was a typical session at the Autodrome that day, with the exception of Ferrari's appearance, a rare occasion. The day's work involved long and—ideally—uneventful hours at the track. It was all trial and error and it required patience. Cars were far less durable and way more idiosyncratic than they are today.

The 330 P was still in its infancy, and the work was dangerous. But the car was performing beautifully. Precise steering. A massive reservoir of power. Every part had been subjected to crack tests under ultraviolet light. Machinists had studied the gear teeth under a microscope for defects. Even the bolts were heat-treated. Surtees lapped for a while then brought the car into the pit to discuss his impressions with the technicians. The computer equipment that would soon revolutionize racing-car development had not yet arrived. Surtees *was* the computer. Gripping the wheel he flew around the circuit, using all five senses to examine the car's behavior.

The threat of Ford Motor Company and the Shelby Cobras didn't register with Ferrari. Though his Formula One campaign had seen its ups and downs, his domination of sports car racing—the kind that translated directly to customer orders—was long-lived and complete. Since 1953, Ferrari had won nine of eleven World Sports Car Championships. Ferrari cars had won Le Mans in 1958, 1960, 1961, 1962, and 1963. The prior year, a Ferrari had finished over 240 miles ahead of the nearest competitor after 24 hours. This new *barchetta*—the 330 P—added 60 horsepower to the previous year's car, while shaving off about five kilograms.

Ferrari believed his new Le Mans weapon would be, when ready, the fastest and most durable racing sports car on earth. In Surtees's hands, it was coming closer and closer to perfection.

THE FORD GT40

JANUARY–APRIL 1964

HENRY FORD II took the stage at Cobo Hall, a downtown convention center on the Detroit River. He looked out and saw countless faces through a haze of cigarette smoke. It was the annual meeting of the Society of Automotive Engineers, and Henry II was to deliver the keynote speech.

"This is a year like other years," he began, "in which America will arrive at a series of crossroads in the long journey in search of our national destiny. For generations our technology has been the most advanced and progressive in the world. It is the basis of our standard of living, of our national security, of our position in the community of nations. We have grown accustomed to thinking of ourselves as the unchallenged masters of the machine age.

"Fifteen years ago, such self-confidence was fully justified. The rest of the world came here to learn how to make things better and cheaper and faster. More recently, in industry after industry, we have seen new processes developed abroad and then adopted here. We have seen foreign products challenging our own, even in our domestic market, not only because the foreign products are cheaper but also because they are often better and more advanced. Signs such as these suggest that we should be asking ourselves some important questions.

"Do we still have more to teach the world than it has to teach us?"

Six time zones away, in a small industrial space in Slough, a suburb of London, a prototype Ford Le Mans car was beginning to take shape. The space was situated in a newly constructed brick factory complex just west of London's Heathrow Airport—convenient for moving parts and people all over Europe and back to Dearborn.

Ford had pieced together a team of brains. Roy Lunn headed up engineering. Ford needed a manager to lead the operation, a man with experience who could be trusted with a checkbook. Shelby brought in his old boss from Aston Martin, John Wyer. ("Pappy can tell you everything you want to know about racing and sports cars," Shelby said by way of introduction.) Wyer was fifty-three, a towering figure with a haggard, sleepless look. His gaze was so fierce, racers liked to call him "Death Ray"—though never to his face. Wyer was Aston Martin team manager when Shelby won Le Mans in 1959. Now, four years later, he signed with Ford for more than twice what he was making at Aston Martin.

Don Frey met with a laundry list of the brightest minds in England in search of an engineering consultant. He found his man in Eric Broadley, an ex-architect in his mid-thirties who'd founded a racing car company called Lola, named for the 1955 hit song "Whatever Lola Wants, Lola Gets." Broadley had shown up at Le Mans in 1963 with a low-slung, midengined prototype racer exactly like the one Ford intended to build. Shelby saw the car and alerted his Ford contacts. Broadley needed money. He could provide experience and contacts to suppliers. Ford signed him to a two-year contract. Shelby's chief engineer Phil Remington rounded out the team.

Already, a flurry of publicity placed a glaring spotlight on the activity at Slough. Le Mans was some 250 days away. Twelve-hour workdays were the norm. Weekends and holidays did not exist. It was only a matter of time before tempers flared. Within weeks, Lunn and Broadley could barely speak to each other. Don Frey was receiving letters from Wyer, second-guessing Lunn's decisions.

Lunn traveled through Europe purchasing the finest state-of-the-art components, none of which were available in America. Borrani fifteen-inch, wire-spoked wheels and Colotti transaxles from Italy. Metalastic driveshaft couplings and Girling brake calipers from England. Other parts came from the engineering laboratories in

Dearborn. All these components needed to be married seamlessly so the whole was greater than the sum of its parts. The failure of any one would result in a Did Not Finish. According to Le Mans rules, no major component could be replaced once the race began.

The first task was to test these components. The team used two of Broadley's Lola cars as mules. Mechanics mounted components at the shop in Slough and trailered the Lolas to racetracks in England and Italy.

It was John Wyer's job to bring in driving talent. He hired a test driver, the best in the business. Today, the name "McLaren" means racing excellence. At that time, Bruce McLaren was a fresh-faced New Zealander with a huge smile and one leg longer than the other. He wore corrective footwear, but he walked with a limp when wearing his racing shoes. As a boy growing up in Auckland, McLaren contracted Perthes disease, a condition of the bones. He spent two years at the Wilson Home for Crippled Children strapped to a "Bradshaw Frame"—a flat bed on wheelchair wheels—with weights hanging off his feet. It was on this wheeled contraption that he began racing. At the Wilson Home there were other kids his age on Bradshaw Frames.

Now twenty-six, McLaren had become a top-rate engineer and a champion pilot. He was the first driver associated with Ford's Le Mans project.

All the car's elements were in design at once. In Ford's styling studio in Dearborn, a team under the direction of design chief Eugene Bordinat had already completed a three-eighths-scale clay model of the body, according to specifications Lunn supplied. This automobile would resemble nothing ever dreamed up in an American design studio.* According to an analysis of the components Lunn had chosen, designers conceived of a silhouette 156 inches long and just 40 inches high, no taller than a kitchen counter. Thus the car's nickname. It was officially named Ford GT (a confusing

* The only American car like it ever built was a one-off research project Roy Lunn created in 1961 called Mustang I (which technically, aside from its logo, had little relation to the Ford Mustang). This midengined car sits today in the Henry Ford Museum in Dearborn.

name, since the car would race as a prototype, not in the GT class). But it came to be known as GT40. Designers painted the clay model blue and white—American racing colors—and shipped it to the University of Maryland for wind-tunnel testing.

As for the engine, the obvious choice was the modified 256-cubic-inch Fairlane V8 that the company was developing for the Ford Indianapolis racing car. Tuned to run for 24 hours, the engine could produce 350 horsepower. That it was modified from the stock Fairlane engine would make for strong advertising copy when the company won Le Mans. The champion car would be, as far as the public relations department was concerned, closely allied to a production Ford.

Before the first car was built, Lunn took the data from wind-tunnel testing and an estimate of the car's weight and horsepower, and plotted it on a graph. The coordinates showed a top speed of 210 mph—faster than anything in Ferrari's arsenal.

The Ford engine produced enough power. All the other elements had to work together to harness it and put it in the hands of the driver.

In Dearborn, Ford engineers were at work on a new suspension system that could survive the brutal abuse of Europe's racetracks while planting all the engine's power to the pavement. Like the human knee, the joint between the wheel and the car's frame is a complex system of structural elements joined by mechanical tendons and ligaments. To create the GT40's independent suspension, engineers used a tool that had never been employed in car design—the computer—plotting movement in different planes and on canted axes.

Electronics would prove critical. As a street-legal car, the GT40 needed blinkers, windshield wipers, an ignition system. Unlike in a production car, the circuitry would be subject to extreme heat and vibration over long periods of time. Rules dictated that the vehicle must be shut down during pit stops and then restarted off the battery, and the car would be disqualified if any lighting equipment failed at night. The team designed the car with two rubber stomach-like fuel bladders wrapped in fire-proof neoprene built into the frame below each door, each with its own filler cap and electric fuel pump. Total capacity: forty-two gallons, the limit according to Le Mans rules.

At Slough, Lunn spent most of his time in the drafting room, where a diagram of the car was sketched across a wall. He knew the transmission was going to be the key. The only one immediately available that met his needs was from Colotti, a small company based in Ferrari's hometown of Modena. It was a four speed. Would it hold up? Drivers would shift some nine thousand times over 24 hours, high-rpm shifts that would place massive loads on the gear teeth. And the disc brakes. Would they last? At the end of the Mulsanne Straight at Le Mans—the fastest stretch of road ever incorporated into a closed racing circuit—the Ford GT40 was going to decelerate from 200 mph to 35 mph in a matter of yards. It was the single most punishing corner on brakes in all of racing.

In mid-December, everything was in place and the team began to build the first car. The frame was critical in determining how the automobile would perform under extreme abuse. Racing engineers in Europe were experimenting with exotic lightweight metals, but there was no time for experimentation. The team chose old-fashioned sheet steel of 0.024- to 0.028-inch thickness. They mounted the frame onto the wire-spoked wheels and positioned the car's heart and soul, its high-revving V8 engine, behind the cockpit and ahead of the rear axle. The skin arrived from the supplier, Specialised Mouldings Ltd., a local company that molded the fiberglass according to a full-size mock-up shipped from Dearborn's styling studio.

The days ticked by, and the deadline loomed. In mid-January, Lunn sent a report to Dearborn. The news was not good. He admitted that the prototype was "well behind schedule" and attributed the tardiness to "the human nature aspects of forming a new team." They were at each other's throats and running out of workdays. Lunn noted that "build time will now run concurrently with race preparation."

The first GT40 was completed on April 1, "eleven months after putting pencil to paper," in Lunn's words. Every square inch was designed to cut through air—the long, sloped nose; rear stance; raised, clipped tail. It was painted navy blue and white, the paint on the hood a matte finish so as not to cause glare. The driver's seat was on the right side, Euro-style. Toggles and gauges were pointed directly at the driver on the instrument panel in various shapes so he could identify them without taking his eyes off the road. The engine sat deep in the chassis, just behind the two seats. On both sides of the car, "FORD" was painted in big letters.

But the work had only begun. The goal was to build two more cars in time for Le Mans. And who knew how the machine was going to behave at speed? A new racing car required extensive development so it could evolve into a safe and efficient machine. As test driver McLaren put it: "A racing car chassis is like a piano. You can make something that looks right with all the wires, the right length, the right size, and pretty close to the right settings, but until it is tuned it won't play so well."

Le Mans practice days, when the GT40 would have its first run at full speed in front of the Ferrari team and an army of writers and photographers, was three weeks away. Meanwhile, the team was down a man; Eric Broadley left. So much for the two-year contract. More bad news arrived.

"Roy?"

"Yes," Lunn said into his telephone.

It was Don Frey calling from Dearborn. Frey wanted the finished car shipped to New York for a press conference on April 3. Both Lunn and Wyer protested furiously. They needed time to put miles on the car, troubleshoot, and adjust. They had no choice, Frey said. Iacocca and Mr. Ford were going to attend the unveiling.

Wyer made the arrangements. The car was flown from Heathrow at 3:00 P.M. on April 2, arriving at JFK Airport in New York. After clearing customs, it was trailered to a Ford dealership for a polish, then to the Essex House on Central Park South in Manhattan. It would leave JFK for London at 3:00 P.M. the following day.

The unveiling was timed for the opening day of the New York Auto Show. Which meant the entire automotive world was on hand. Sitting on a carpeted floor in front of a curtain at the Essex House sat perhaps the most expensive marketing tool in history. It looked like something out of the James Bond movie *From Russia with Love,* which premiered in New York that same week.

Here was an American sports car that could travel 95 mph—in reverse.

When Mr. Ford walked into a room, executives straightened their spines. Henry II looked down at the car for the first time. He saw a sexy, low-slung rocket of a European racing car with his name on it. He wondered what his stingy grandfather would have thought of the GT40. Henry Ford probably would have hated it. But Henry II's father Edsel would have loved the thing.

A crowd filled the room, and Ford executives mingled. They were all here: Frey, Lunn, Wyer, Iacocca. Photographers snapped shots of the car, the camera flashes sparkling in its wire-spoked wheels. Reporters jotted impressions in their notebooks. There was an air of unconcealed excitement. A Ford car to race against the Ferraris at Le Mans—those reporters couldn't have scripted a more enticing plot.

Iacocca took to the podium and delivered a short speech, with talk of 200 miles an hour and 350 horsepower. The engine was built in the United States, the brakes were from England, and the transmission was from Italy. This was not just an American racing car, it was *The World Car,* an embodiment not of a Detroit company but a global empire. As Iacocca spoke, Lunn and Wyer sat in the audience listening. They feared the worst. Le Mans practice days were two and a half weeks away and their car had barely turned a wheel.

A few days later, Wyer stood in the pits at Goodwood, a racetrack near England's southern coast, watching the Ford GT40 as McLaren looped it around the winding circuit. Wyer instructed McLaren to drive "at medium speeds," no faster than 145 mph. A couple of short shakedowns were all they could do to develop the car before heading to France. Standing next to Wyer in a leather jacket, leaning up against the pit wall with hands folded in a ball, stood Phil Hill. Except for a female assistant holding a stopwatch, the two men were alone. The grandstands stood empty.

That spring, Hill was trying to find himself. He'd spent the 1963 season with an upstart Italian racing firm called ATS, but the company was a failure and so was Hill. He didn't finish a single F1 race. Pundits said the American champion had lost his confidence. Worse still, some said he had lost his courage, his nerve cracked. There could be no more damning criticism for a man in Hill's shoes.

The former champion found himself without a Le Mans ride for 1964, and Wyer signed him to lead the American effort along with McLaren. As Hill watched McLaren cruise past the pits in the Ford car, the irony must have burned into him. How quickly his fortunes had turned. The press had called him "Mickey Mantle in a Ferrari" two and a half years earlier. Now he was "Hamlet in goggles and gloves."

He was to take on his former boss Enzo Ferrari at Le Mans. Was Hill in fact washed up? The most accomplished endurance driver in the world was the man who had the most to prove.

LOSS OF INNOCENCE

APRIL–JUNE 1964

> In the long run, death is the odds-on favorite.
>
> —EDDIE SACHS

ON THE MORNING of April 18, 1964, a group of mechanics pushed two Ford GT40s down a muddy path at the Le Mans circuit, underneath the grandstands to the pit lane. Roy Lunn and John Wyer walked alongside soberly, dirtying their shoes. According to Lunn, the two Fords had "an aggregate of only four hours running time with no high-speed experience." They were geared to run at 200 mph down the Mulsanne Straight, faster than anything had ever traveled at Le Mans before. But who knew what they would do in actuality?

The Le Mans test weekend was an institution. The race was held on public roads that, come June, were cordoned off and turned into a closed racing circuit. So each April the organizing committee, the Automobile Club de l'Ouest, sealed off the circuit so teams could test the new machinery at speed without having to dodge car and tractor traffic. Conditions were typical for spring that morning: a sky so overcast it appeared almost purple, with a misty drizzle that made for slippery pavement. Spectators and journalists were given free access to the pits and they crowded around the mysterious Ford racing cars.

Soon cars were speeding down the straight past the pits, kicking up wakes of rainwater, windshield wipers thumping. Ford's top drivers were competing that weekend in the Aintree 200 in England. Hill and McLaren were contractually obligated to race for the Cooper team; they were planning on flying into Le Mans to turn some laps in the new cars before the weekend was over. Wyer had brought in two other drivers—Roy Salvadori and Jo Schlesser. Salvadori was a Le Mans champion, having won in 1959 driving for Wyer in an Aston Martin, teamed with Shelby. Schlesser didn't have the same pedigree. He was co-owner of a Ford dealership in Paris and was more known for crashing cars than winning. Ford of France was keen on having a Frenchman on the team. The Paris office wanted a piece of the action in more ways than one.

Salvadori was first onto the track. Rain-slicked gray helmet on, he pulled his leather gloves over his hands. When he fired the engine, its aria shrieked from the dual crossover tuned exhausts. He put the car in gear, rode the clutch for a moment, and the Dunlop tires gripped the wet pavement. As the car eased onto the straight Lunn stood beside the track wringing his hands nervously. He wiped his nose with his sleeve and watched the car pull away.

In the pit Wyer and Lunn waited, eyeing stopwatches as the Ford made its maiden voyage around the historic 8.36 miles. Up under the Dunlop Bridge, down through the Esses, along the 3.5-mile Mulsanne Straight. Minutes later, Salvadori appeared again, rounding the White House bend at an easy pace. He pulled back into the pit, screeched to a stop, and got out of the car. From the look on his face, everyone in the pit could tell he wasn't happy.

"I can't believe this, John," Salvadori said to Wyer, "but I think we're getting rear wheel spin at 170 mph."

Lunn and Wyer conferred. This was incredible—at that speed the rear of the car was lifting off the ground. An argument broke out. Was the problem aerodynamics, or a suspension issue? The latter idea won out and mechanics went to work, making adjustments. Salvadori was spooked; he wanted no part of this experiment and was done for the day.

Schlesser strapped on his helmet. Wyer leaned in and gave the French driver instructions, likely saying, "Do not take chances.

Bring the car back in one piece." Soon Schlesser was off, accelerating under the Dunlop Bridge and out of sight. In the pit the Ford team waited. A little over four minutes passed and Schlesser appeared, motoring out of the White House bend and past the empty grandstands. He was moving quickly, the V8 engine cranking power through fourth gear to the rear wheels. Again Schlesser moved under the Dunlop Bridge and the Ford disappeared. A minute passed. Then another. Eyes craned toward White House, waiting for Schlesser to turn the corner. Another minute passed. And another.

The car never appeared.

A phone call came in from the signaling pit on the far side of the track by the end of the Mulsanne Straight. There had been a crash. It appeared quite devastating, but word was, the driver was alive. When Schlesser appeared in the Ford pit, he was shaken and bleeding from a small gash on his forehead. He'd caught a ride back.

The car was fishtailing all the way down the straight, he complained with a French accent. "It wouldn't go in a straight line." Schlesser had been traveling at roughly 160 mph when he lost control. He was in need of a brandy and a chair.

For Ford, the test day was over. As sports cars painted in various national colors looped around the circuit, Lunn and Wyer walked by back roads and through wet fields to see what was left of their car. The crash site was in a woody area at a kink in the Mulsanne Straight called *La Grande Courbe*. It was a dangerous area prohibited to spectators, but the two men pushed through. Standing there, they eyed what was left of the GT40. It had taken them months to build and now here it was, in pieces scattered along 300 feet of the roadside. The car was totaled.

"It is incredible that he escaped with his life," Wyer said.

The next day, Salvadori crashed the only other GT40 in existence. He was not injured, but the car was. A Ford rep called back to Dearborn and the phone rang in Don Frey's office. "We wrecked both of them," the rep told Frey. "I'm up to my hips in shattered cars."

When the weekend ended, the team gathered the broken machinery and headed back to England. Long hours and sleepless nights awaited. Le Mans was two months away, and the test weekend had raised more questions than it had answered. The *New York Times* on Monday morning: "In trials that ended yesterday, a Ferrari car driven by John Surtees was clocked at 194 miles per hour down the three-mile Mulsanne Straight. Nothing has ever traveled here that fast before. And two Ford prototypes crashed. These Fords were new, unbelievably sleek and expensive . . . People who know money think Ford can build a winner. People who know car racing are not so sure."

Lucky for Lunn and Wyer, Henry II was otherwise occupied. That same weekend, he and Iacocca flew to New York to unveil a new model at the World's Fair. It was the Mustang. If the Ford Le Mans car was a marketing tool, this was its raison d'etre.

Henry II and Iacocca chose the opening weekend of the World's Fair to lure maximum exposure. The company's exhibit—officially called "The Wonder Rotunda," largely designed by Walt Disney—sprawled over seven acres next to the brand-new Shea Stadium, home of the New York Mets.

Iacocca took to the stage clad in an Italian-cut gray suit. He stood next to his car, the first model he could call his own.

"We appreciate your coming here to share this moment with us, one of the most important occasions in Ford division history," he told a bustling crowd. "Incidentally, this is Ford's first international press introduction. While we meet here, the Mustang is being introduced in 11 European capitals to some 2,000 reporters, editors, and photographers."

Iacocca was spending more than $10 million on a tsunami of media in an attempt to instantly embed the Mustang in the public consciousness. The following Thursday, the company had half-hour programs running in prime time simultaneously on all three major television networks—a first in the annals of broadcasting. Iacocca's face graced the cover of both *Time* and *Newsweek*. He tricked both weeklies into thinking they had exclusives, and both made him the cover story, calling him the man behind "a new breed out of Detroit." He'd scored the publicity coup of the decade. Suddenly everyone knew how to pronounce the name "Iacocca." (*Time:* "Rhymes with try-a-coke-ah." *Newsweek:* "Pronounced eye-uh-coke-uh.")

The Mustang was a winner—there was no question. It captured a moment in time, the year 1964. American attitude coupled with European style, horsepower as a metaphor for youth and modernity. It was all here in this one automobile. Sticker: $2,368 plus tax. For an extra $437.80, buyers got a 260-cubic-inch V8 engine, a whole lot of muscle. Simulated knockoff European-style racing hubcaps: $18.20. Racing package with stabilizer bar and stiffer struts: $38.60. *Time* waxed thick about the new Ford: "With its long hood and short rear deck, its Ferrari flare and openmouthed air scoop, the Mustang resembles the European racing cars that American sports-car buffs find so appealing."

The media blitz shocked Henry II. He was stunned by Iacocca's shameless self-promotion. Those *Time* and *Newsweek* stories barely mentioned Henry Ford II. Publicly, Henry II could only praise his employee. Privately, he seethed with discontent. Iacocca's delight in the pages of *Time* was both portentous and naïve: "I see this as the start of a new golden age for Ford that will make the peaks of the past look like anthills."

Days later, on the eve of Memorial Day, Iacocca had dinner with a racing driver named Eddie Sachs in Indianapolis. It was the night before the Indianapolis 500, and Sachs was going to be driving a Ford-engined car in the race. The driver and Iacocca were from the same town—Allentown, Pennsylvania—and they'd gotten to know each other. At thirty-seven, Sachs was Indy's favorite underdog. He'd started out as a dishwasher in the Indianapolis Motor Speedway cafeteria, and he was obsessed with winning the Indy 500. He'd been hospitalized thirteen times over his career. The left side of his face had been reconstructed by plastic surgeons. Iacocca would always remember how calm Sachs was at dinner that night. He seemed perfectly at ease.

The next day, Iacocca took his seat in the "Penthouse Paddock" at the track along with an entourage of Ford executives. He watched the Purdue University marching band play "Back Home Again in Indiana." Track owner Tony Hulman made his famous announcement over the loudspeakers:

"Gentlemen, start your engines!"

Iacocca's pulse quickened when he saw the pace car lead the thirty-three racers around the track in front of 250,000 spectators. The pace car was a Ford Mustang convertible, with Henry II's younger brother Benson at the wheel.

When the Mustang pulled into the pit lane and the green flag waved, Iacocca felt the explosion of the engines like a kick in the sternum. He watched the cars loop around the track into lap two. On turn four, an Indy rookie named Dave "Mr. Sideways" MacDonald lost control of a Ford-engined car and smacked its nose hard into the retaining wall. The car bounced back into the middle of the track. It happened so fast and early in the race, the sound of the screeching tires took the crowd by surprise. Coming around turn four at full speed came Eddie Sachs. He had nowhere to go. Sachs steered his machine, loaded with a full tank, broadside into MacDonald at 150 mph. A deafening blast released a burst of flames and a black mushroom cloud. Exclamations poured from the crowd.

"Jesus Christ! Lookit that smoke!"

"It looks like an atomic bomb!"

All over the country, closed-circuit television screens went blank. In the Penthouse Paddock, Ford executives gazed down at the smoldering wreck. Two cars with "Powered by FORD" painted on them had turned into blazing coffins in front of hundreds of thousands of spectators. Iacocca's face went slack. The flames burned holes into his dark brown eyes.

LE MANS, 1964

> I am hypnotized by the atmosphere: the trees, the immense stands, the undulating roads with such holy monsters as Tertre Rouge, Maison Blanche, and the Hunaudières straights, the legendary Arnage and Mulsanne corners, where many dreams have foundered and others have brightened. I stand there for a half hour looking around and I relive the noise of the people, the roar of the crowds, the exhaust notes of the cars, the frenzy of refueling, the lights at night, the livid light of dawn, the hot sun that worsens fatigue and mists the mind. Ah, the stress of Le Mans!
>
> —Enzo Ferrari's lieutenant FRANCO GOZZI

NO ONE BELIEVED the Americans stood a chance. It would be a miracle if they beat the Ferraris in their debut at Le Mans. In fact, it would seem a miracle if they could keep their racing cars on the road. But then, in the spring of 1964, people had grown used to the unexpected, to heroic events and shocking headlines. In the previous twelve months, John F. Kennedy had been assassinated, the U.S. Congress had passed the first Civil Rights bill, and the Russians had launched the first woman into space. A twenty-two-year-old Louisville heavyweight named Cassius Clay had knocked out Sonny Liston in Miami Beach. Martin Luther King had marched on Washington and had delivered a speech that changed everything. "I have a dream," King chanted, and so many other Americans had one, too.

When the Ford team checked into the Hotel de France in La Chartre-sur-le-Loir, a twenty-mile drive from the Le Mans circuit, chaos was already waiting for them. The Hotel de France was John Wyer's spot. It had served as the Aston Martin Le Mans clubhouse through the 1950s. In 1950, Wyer housed the entire Aston team for nine days at the hotel and the bill came to 462 pounds sterling. In 1964, he couldn't buy the Ford team lunch for that amount.

An army was arriving from Dearborn. Always, the week leading up to the 24-hour race was a frenzy of activity, but this was absurd. Carburetor specialists, tire and engine men. Wyer was in charge, but he'd never seen many of these faces. There weren't enough rooms to go around. The place was crawling with reporters. A transporter carried three prototypes to a guarded paddock beside the hotel, where mechanics began all-night sessions taking the cars apart, inspecting every piece, and rebuilding them. The team had made significant modifications since the disastrous test weekend two months before, most notably a rear spoiler. The three-inch piece of metal jutting off the car's tail created down force on the rear end. The spoiler "had the effect of putting feathers on an arrow," as Lunn put it.

Wednesday through Friday were practice and qualifying days, and the race started at 4:00 P.M. on Saturday. It all had to go like clockwork, down to the customs papers to get the Fords into the country. The cars had to withstand a rigorous inspection; the French officials were legendary for their fastidious attention to an endless list of seemingly random regulations.

Wyer ran the show like a general. He had a stomach ulcer and was perpetually in a bad mood. He set up a schedule and the army had to live by it. When one member of the public relations team went in search of lunch at the hotel at 1:00 P.M., he was told, "Mr. Wyer doesn't have lunch until 4:00 P.M. No one can have lunch until 4:00 P.M." Wyer posted timetables for mechanics—which men would attend track sessions, which would work the overnight shift. When one team of mechanics finished their shift, the next would report for duty having slept, showered, and shaved.

The Dearborn suits were on edge before they arrived at Le Mans. There was hell to pay for the deaths of Eddie Sachs and Dave MacDonald at Indy three weeks earlier. A full investigation was underway. Ford's PR chief made a statement: "We are all shocked and saddened by this tragedy. But I don't think it should be a factor in making us pull out of racing. It's dreadful that it happened. But this is built into racing."

On the morning of the first Le Mans practice session, the pit lane filled with red Alfa Romeo Giulia TZs, silver Porsche 904s, green Triumph Spitfires and Jaguar E-Types. John Surtees was spotted, as was Phil Hill. Carroll Shelby arrived with a pair of Cobra Daytona coupes, painted guardsman blue with white stripes. There was no way to measure the man hours, ingenuity, loving care, and soul that had gone into these cars. Shelby was a fan favorite in France. When he walked out onto the pavement and looked up at the empty grandstands towering high, it all came back to him: the *magic* of this place. He was a champion here; no one could ever take that from him. Still, it must have felt like a different lifetime. He was a constructor now, armed with Ford money. If his Cobras could win the GT class—something no American car had ever accomplished—his little automobile company would be assured survival.

"Outside of the United States," he told a *Sports Illustrated* reporter, "the Le Mans race has more prestige than all the other races put together. Le Mans receives throughout the world probably five times as much publicity as Indianapolis. Any automobile manufacturer who wants to make a name for himself in racing has to do well at Le Mans."

The first engine sounded and soon revs were coming from all directions. The air stank of exhaust and hot pavement. One by one, cars motored onto the circuit. Stopwatches clicked off vital seconds. The press box grew loud with the sound of thumping typewriters. Facing the three Fords and two Cobras, Ferrari had entered four cars and a number of privateers were racing their own Ferraris, also prepared at the factory by Ferrari's men, bringing the total to eight entries branded with the Prancing Horse.

From the first day of practice it became apparent that the race would move at historic speeds. One after another, Ferraris cut deeper into the circuit, shattering the Le Mans lap record: 3:47.2. Then 3:47. By the end of qualifying, the crowds that had begun to amass were left with a cliffhanger. Surtees set the best time: 3:42. His speed was dumbfounding. He'd knocked more than ten seconds off his own lap record from the year before. But a Ford qualified next, in the hands of Californian Richie Ginther. The Mexican Pedro Rodriguez qualified third and Phil Hill was fourth. Over the 8.36-mile course, less than four seconds separated the top four qualifiers.

On the eve of the race, Surtees stood in the Ferrari garage taping an on-camera interview with Stirling Moss for ABC's *Wide World of Sports*. Until three years earlier, Moss had been considered the greatest racing driver in the world. Some said he was the best who'd ever lived. One high-speed injury later and here he was, with a microphone in his hand rather than a steering wheel. Interviews were not Surtees's forte. The camera made him more nervous than a dice on the Mulsanne Straight. Moss asked him about the American threat. How important was it to Enzo Ferrari to beat the Fords?

"To a firm like Ferrari," Surtees said, "which produces a specialized product and sells most of its cars in America, it's very important."

"Ferrari has won this race four times in a row," Moss said, "and if he wins this race it'll be five times, which has never been done. You're entering four cars?"

"Yes."

"How many men did you bring?"

"Our team is comprised of about twelve or thirteen mechanics, one engineer, and one team manager."

Moss looked around the garage. There were seven cars. "What are the extra cars for?"

"In case anything unusual happens," Surtees said. "For instance, the other night we were out and we hit a fox in the middle of the road at about 140 mph. It could have damaged the car rather badly."

"Well I imagine it damaged the fox rather badly," Moss laughed.

A smile crept out of the side of Surtees's face.

Behind him, the cars were lined up in a row on the cement floor. Mechanics in beige jumpsuits took a break from wiping them down so they could leer at ABC's script girl holding cue cards near the camera. A handful of men in suits wandered about, everyone with hands in pockets, resisting the urge to get fingerprints on the red metal. An air of complete confidence permeated the garage, as if the Americans posed no threat whatsoever.

"After all," joked Luigi Chinetti, "the best American sports car is the Jeep, no?"

When Surtees wrapped his interview, he started to think about sleep. With Ferrari, there were no dramatic meetings, no strategies to coordinate. Out on the track it was every man for himself. Surtees was teamed with Lorenzo Bandini, Ferrari's number-two driver. Together they were the odds-on favorites.

At the Hotel de France, Wyer assembled the Ford racers for a meeting. Six drivers, three teams of two. Wyer's philosophy was the opposite of Enzo Ferrari's. He believed in a team approach. Each driver and car was a cog in his victory machine. He wanted everything done precisely according to his orders.

"We want to finish the race," Wyer said. "We aim to keep our cars running. We all must remember, this is an endurance race, not a sprint race." Phil Hill and Bruce McLaren, Ford's two superstars, comprised the number-one team. Wyer's master plan had them winning. They would keep pace with the front-runners. "Stay close at court," Wyer ordered. "Speed must be as high as possible while conserving brakes and gearbox. You must stay in a position to strike if attrition takes its toll on the leaders, which it always does." Wyer advised the junior members of the team—Briton Richard Attwood and Frenchman Jo Schlesser—to stay back with the pack and go easy on the car. Their job was simply to finish the race.

Wyer turned to Richie Ginther, a short, toothpick-shaped man with red hair and an impressive résumé. Ginther had raced on the Ferrari Formula One team, and was an old friend of Phil Hill's back to the days when they both had worked at Roger Barlow's automobile dealership in Los Angeles. Ginther had qualified fastest on the Ford team. He was partnered with Masten Gregory, "The Kansas City Flash." Wyer ordered Ginther to run hard at the start to try to get the Ferrari drivers to break their engines.

Ginther got the point. The opening laps would be a dogfight. He was going to show the world what the new Ford could do.

All roads leading into Le Mans clogged with overheating cars, trunks filled with tents, sleeping bags, and Kodak Instamatics. Cabs moved bumper to bumper past the Le Mans train station. By the afternoon, spectators had swamped the grandstands and crowded the fields around the circuit. According to French officials, the largest crowd ever was attending the race, well over 300,000. Strolling through the mobs one heard English, German, Italian. Barracks full of American soldiers in uniform, on leave from base in West Germany, formed lines at the beer tents. The airfield behind the grandstands was busy with traffic, private planes and helicopters carting in bigwigs.

Mechanics began pushing racing cars out of the paddock and onto the pit straight at the bottom of the grandstands. Flowers piled up under a plaque that marked the spot where Pierre Levegh had crashed into the crowd nine years earlier; the plaque read simply, "June 11, 1955." The official Dutray Le Mans clock hung over the pavement in the center of it all, and as its hands rounded closer to 4:00 P.M., drivers appeared holding their helmets. They wore racing shoes, leather gloves, and fireproof coveralls, goggles draped around their necks.

The Le Mans start was foreign to the American racing fans. Drivers stood on one side of the road across from their cars, which lined the pit row in order of qualifying, the fastest car at the front. The starter stood in the center of the road holding the French flag high and when he dropped the flag at exactly 4:00 P.M., the drivers sprinted across the two-lane road, jumped in the cockpits, hit the ignition, and boxed each other into the opening straightaway in the fiercest and loudest traffic jam ever witnessed.

Minutes before 4:00 P.M., gendarmes herded the crowds off the pavement and the drivers took their positions. In Italy, Enzo Ferrari sat down in front of a television. In the pit, Shelby paced; the reputation of his company was on the line. His Cobra had clocked 197 mph in qualifying on the Mulsanne Straight. A host of high-level Ford executives had arrived and they stood in the Ford pit waiting and watching. Following a handful of national anthems, silence settled over the hundreds of thousands of spectators. Smokers could hear the crackle of their cigarettes burning. Rows of photographers lined the pavement aiming like gunners in a firing line. A voice over the loudspeakers counted out the final moments in French.

". . . Thirty seconds . . . ten seconds . . ."

Start

Phil Hill dashed across the road. He jumped through the right-side door into the Ford's cockpit and hit the ignition. The V8 came to life. Clutch in, shift into first, down on the gas, up on the clutch. The engine stalled. Hill saw cars peeling off all around him onto the opening straight. The noise was deafening even through earplugs. And then he was alone on the starting line. He couldn't get the car to move. He couldn't goddamn believe it. In the pit, mechanics and Ford executives looked on, their jaws hitting the pavement. By the time Hill got the car going, he was alone, motoring down the straight in last place, gearshifts crackling in rapid fire.

Even then Hill knew something was off. Something was very wrong.

Surtees tore down the opening straightaway, up the slight right-hand incline, and under the Dunlop Bridge. He loved the pavement at Le Mans—"billiard table smooth." Two other Ferraris got a jump on him and he found himself trailing in third place.

It was a long race.

The early laps were among the most dangerous, when not-so-skilled drivers swapped paint at high speed. Which is why it was wise to motor ahead of the riffraff as soon as possible. Surtees focused on the course, every nerve and reflex in tune with the movement of the car and the other cars around him. He was merciless in close combat. No matter how good you thought you were, he'd find a way to pass you, leave you wondering, your concentration snapped. It was custom for drivers at Le Mans to wait until they reached the Mulsanne Straight to strap on their seatbelts; on the straight they could hold the wheel with their knees.

By the time Surtees was hauling back through the grandstands at the end of the first lap, it was one-two-three for Ferrari. A flagman stood in the center of the lane signaling caution; slick oil had already spilled onto the pavement.

In the cockpit, everything unfolded in slow motion. "When you start [racing]," Surtees once wrote, "120 mph seems like 160 mph. With experience, that 120 mph seems more like 60 mph." As Surtees maneuvered the twisty, downhill Esses on lap two, he saw in his rearview mirror the mouth of a Ford GT40 tuck in behind him. Mere inches separated the two cars. Surtees downshifted into second gear and turned into the right-hand *Tertre Rouge* corner onto the Mulsanne Straight. Then he accelerated hard, with the Ford slipstreaming behind him. Third gear, fourth, fifth. Surtees was approaching 190 mph. The world turned into a Technicolor blur, as if he was being sucked into some cosmic vacuum cleaner.

Suddenly the Ford jumped to the left to pass. It was the #11 car. Richie Ginther darted past Surtees traveling faster than any car ever had on the storied Straight. Surtees saw him through his windscreen and—just like that—Ginther was gone.

In the press box, ABC's Jim McKay was yelling wildly into his microphone, taping footage for the next weekend's *Wide World of Sports* broadcast: "Word from the course is that Richie Ginther, who had moved up from eighth to fourth place, has passed some more cars. As a matter of fact, the word is that Richie Ginther has taken the lead in the second lap in the white Ford with blue stripes. The American racing colors are in the lead at Le Mans! There he is on the right of your screen. Get a look at that low-slung Ford! I've never *seen* a car as low as that!"

Phil Hill was back in the pit and mechanics were digging into the engine compartment. Minutes were speeding by, Hill losing more and more ground. The crew found the problem: a blocked jet in one of the Weber carburetors. The car couldn't breath. These Italian-made carburetors were so complex, the mechanics were dumbfounded by them. Where the hell was the Weber representative? Against the noise of engines John Wyer heard a man shouting at him from above in the crowd. He looked up. It was the Weber rep.

"What are you doing up there?" Wyer yelled.

"They wouldn't give me a pit pass," came the reply.

Not soon enough the carburetor was fixed and Hill raced off in a healthy GT40.

Cramped into that small cockpit, the champion began to weave through the traffic. By this time Hill was in forty-fourth place. He'd lost twenty-two minutes. To catch up to the Ferraris from that distance would require the powers of a superhero. Hill knew this circuit better than any man.

How to Go Fast

Hill began to rip off a series of perfect laps. Experience told him how to make up time at high speed without overtaxing the engine.

Enzo Ferrari, "the Magician of Maranello," standing at the gate of his legendary factory. (*Toscani Archive / Almari Archives / The Image Works*)

Henry Ford II as a young auto man. On the desk is a portrait of his father, Edsel, and his grandfather Henry Ford. (*Gjon Mili / Time Life Pictures / Getty Images*)

Lee Iacocca (right) and his chief engineer, Donald Frey, in 1965 with the new Ford Mustang (right). (*AP Photo*)

Phil Hill (left) with Enzo Ferrari in the pits at Monza before the Italian Grand Prix in 1958, Hill's first race in a Ferrari Formula One car. (*Klemantaski Collection*)

Hill became the first American World Champion at the 1961 Italian Grand Prix, a race that killed fourteen spectators and Hill's friend and rival Count Wolfgang Von Trips. (*Klemantaski Collection*)

Carroll Shelby, on crutches following knee surgery, led his band of hot-rodders to Sebring in March 1964. (*Dave Friedman*)

Englishman John Surtees won seven Grand Prix World Championships before joining the Ferrari team. (*Terry Fincher / Keystone Features / Getty Images*)

Surtees at the Cavallino with Ferrari (right) and Ferrari's lieutenant Franco Gozzi (middle). (*Julius Weitmann / mpi photoservice*)

Ferraris on the track at Le Mans in 1964, the first battle against Ford Motor Company at the world's most famous race. (*Klemantaski Collection*)

Shelby driver and former British tank commander Ken Miles.
(*Dave Friedman*)

Victory at the Daytona Continental in February 1965: from left, Ken Miles, Carroll Shelby, Lloyd Ruby, and Leo Beebe. (*Dave Friedman*)

Carroll Shelby and Phil Hill leaning on a Ford before Le Mans in 1965.
(*Dave Friedman*)

The Ferrari 330 P3, built to battle the Americans at the ultimate racing showdown—Le Mans in 1966. (*Don Heiny / Corbis*)

A replica of the Ford Mk II, Henry Ford II's Le Mans weapon in 1966. (*Courtesy of Ford Motor Company*)

Top: Bruce McLaren, Ford driver and founder of the McLaren racing team.
(*Keystone Features / Getty Images*)

Right: "Newcomer at Dangerous Le Mans" read the *New York Times* headline when Le Mans rookie Mario Andretti joined the Ford team.
(*Robert Riger / Getty Images*)

This photo was taken seconds after the veteran Walt Hansgen crashed his Ford Mk II at the Le Mans trials in 1966. Hansgen was trapped inside and still alive.
(*Dave Friedman*)

"The Last Tycoon," Henry Ford II, arrives at Le Mans in 1966. (*NOA / Roger Viollet / Getty Images*)

The start in 1966. (*Dave Friedman*)

The most controversial finish in Le Mans history. (*Klemantaski Collection*)

There can be only one shortest distance around a racetrack, achieved when the driver chooses the perfect line on every turn. When he moved the car through a bend he could ease the tires within an inch of the pavement's edge. The great endurance racer possessed a kind of compassion for the machine and its countless moving parts, allowing it to breathe and flex its muscle.

In large part, the race was won or lost on the rev counter, the rpm gauge staring the driver in the face from the center of the instrument panel. If Hill aimed to take a turn at 4,500 rpm, 4,400 rpm wasn't good enough. The difference between a 4-minute lap and a 3:58 lap equaled roughly 25 miles at the finish.

In the grandstands, fans watched Hill shriek out of the White House bend and down the pit straight. Thumbs clicked on stopwatches when he flew past the start/finish. He was cruising at 185 mph in fourth gear at 5,700 rpm. A slight inclining right bend led him under the Dunlop Bridge. He eased up on the gas, then accelerated again, shooting down a slope at 183 mph into the Esses. He downshifted to third, then second, in perfect fluid motion. Easy on the downshifts; no stress on the gear teeth or clutch plate. Hill left the Esses in second gear at 5,800 rpm—82 mph. A hard brake down to 65 mph, a tight right turn onto the Mulsanne Straight, and he hammered the throttle. Third, fourth. The g-forces pinned him against his seat. A glance at the tach: 6,100 rpm. Two hundred mph summoned with his toe.

Nearing the end of the Straight, a blind right-hand kink approached—*La Grande Courbe.* Hill took the kink flat out. Then came the Mulsanne Corner, the hardest turn on brakes in racing. He let the car coast . . . Then he nailed the brake pedal and downshifted: three, two, one. Exhaust pipes spit sparks and the cast iron brake discs turned fiery red. The lap belt dug into Hill's waist. He steered into the right-hander at 35 mph.

Hard on the accelerator. Second, third, past the signaling pits on the right, back up to 180 mph. Hill hurled the car through turns, rear wheels struggling for grip. The grandstands appeared in the distance. Hill gunned through that chasm, a huge valley coated with human bodies. Many thousands of eyes followed the blue and white streak as it passed the start/finish, a Ford car hurtling 185 mph on four patches of rubber. Thumbs clicked on stopwatches again. The American was moving.

No two laps were the same. Hill's brain filtered stimuli, automatically ranking it in order of importance in nanoseconds. Photographers leaning in and waving at him. A piece of newspaper floating in the wind. Pit signals, which might communicate a driver's position, his lap speed, or a signal to bring the car in for a stop. With each lap, fuel burned off, lightening the car, increasing its speed. Perception was near extrasensory. "True concentration is not aware of itself," Hill explains. When the driver forgets himself is when he becomes one with the machine. The engine is ignored unless there is a problem. The heart is racing, maybe 160 beats a minute, but it, too, is ignored. "The flagmen, unless they are holding a yellow flag or some such thing, are perceived and forgotten," Hill says. "A small car you are overtaking is registered and erased as you safely pass." Always there is the danger that a driver in a car moving 80 mph slower will make an unexpected move.

As Hill weaved through the field, the cockpit heated up. During daylight hours it could hit 140 degrees Fahrenheit. Dressed in coveralls, helmet tight over the head, the body began to dehydrate. Noise numbed the ears and the same brutal, incessant vibration that threatened the car's electronics wore on the driver's nervous system. Lap after lap, hour after hour. "Sometimes you may not even be aware of the break in your concentration," Hill says. "Not until you find yourself plunging past your braking point."

Pit Stop

Richie Ginther pulled his #11 Ford into the pit. It was just after 5:30 P.M. Ginther stepped out of the car and the crowd roared for him. He was in first place.

None of the mechanics said anything. Four of them—the most allowed by Le Mans regulations—went to work. They were wearing armbands with *MECANICIEN 1964* printed in black so they were easily identified by the hovering officials. Tires to check, tank to fill.

"Well, for God's sake," Ginther shouted, "isn't anyone going to ask me how the car went?"

Questions followed and Ginther told his story. One man present described him as "wildly ecstatic." When he passed those Ferraris to take the lead on the Mulsanne Straight, Ginther said, his tach read 7,200 rpm. He had hit 210 mph. Only two months earlier, Surtees had set a mark on the Straight during the Le Mans test weekend at 194.

Ginther's teammate Masten Gregory hustled over to the car. He looked over his shoulder at his boss John Wyer, nodded, then jumped in. But the mechanics were not finished. The whole team watched and waited. And waited. No matter how fast the car traveled, it meant nothing if pit stops were slow. By the time the #11 Ford screeched onto the pit straight, two minutes seven seconds had passed.

Surtees had taken the lead.

Attrition

In the Cobra pit, Shelby stood making a meal of his fingernails. At 9:00 P.M., one of his Cobras was leading the GT class in the hands of Dan Gurney, miles ahead of the Ferrari GTOs, lying fourth overall. Gurney had raced here six times, but he'd never finished. He had a heavy foot, perhaps too heavy for this race. The Cobra had a 5-mph edge in top speed over the Ferrari GTOs, but those Ferraris were bulletproof. As one GTO pilot put it: "A Ferrari was like insurance. You were assured that you would finish the race."

Would Shelby's Cobra hold together? The tall Texan rubbed his eyes and watched the Cobra as it passed, as if the intensity of his stare could somehow ward off mechanical failure. The sun ducked slowly behind the grandstands.

John Wyer's careful plans began to unravel. A little more than four hours into the race, the team received word that a GT40 had burst into flames on the Mulsanne Straight. Word from the signaling pit on the other side of the circuit: the driver, Richard Attwood, had climbed safely out of the car, but it was still burning on the side of the road.

When Attwood made it back to the pit he saw Wyer's sour face. "I saw some flames coming up around the induction system in my rearview mirror," the driver said, explaining that he parked the car on the side of the road and jumped out. "I literally watched the fire take hold of the car. It seemed like a long time before any attention came to put the fire out."

One of the three prototype Fords was retired. Wyer later learned that the fuel hoses, which were supposed to be made of an ultra-durable synthetic material, had in fact been made of plain nylon, and the heat of the engine compartment had melted one of Attwood's hoses. "This was the result of almost criminal negligence," Wyer later commented. "It was a miracle the other cars were not affected."

Meanwhile, Masten Gregory was battling with Surtees for first place. The announcer's voice in French over the loudspeakers signaled that something had happened. When Gregory plunged past the grandstands again, he was leading Surtees. Morale in the Ford pit soared. Don Frey leaned in to Wyer.

"It is enough," Frey said. "If we do nothing more in this race, I am satisfied."

Ha, Wyer thought, *I should get that in writing.*

Minutes later, Gregory limped into the pit. He was having trouble with the transmission. He couldn't get out of second gear. Mechanics went to work but it was futile. Wyer gave word to the officials; he was withdrawing a second Ford. Roy Lunn stood by, the irony eating at him. The faulty component—the Colotti transaxle—was made in Modena, Italy, by a former Ferrari employee.

Only one Ford remained. Hill was still far behind the leaders, with 19 hours to go.

Night

After sunset, spectators no longer saw the silhouettes of cars on the track but rather headlights stabbing through the dark. Speeding shadows could be identified not by shape and color but by exhaust note. Keen ears could pluck out the song of the Iso Rivolta, the Porsche 904, the thunderous GT40.

Darkness added an element of danger. To aid vision on the Mulsanne Straight, tree trunks were painted white so they would reflect headlights. Some drivers preferred the action after dark. "Driving at night, once you become accustomed to it, you find that the very high speed is much safer than during the hours of daylight," Bruce McLaren later wrote in his diary. "The main danger at Le Mans was the little cars with a top speed around 90 mph that were cruising nearly 100 mph slower than we were, but in the darkness they couldn't help but see our lights coming up behind, and they stayed out of our way."

McLaren took over for Hill at midnight. He later described this four-hour shift as "the best 500 racing miles I've ever covered."

For the crowd, the party picked up steam. From its inception, Le Mans had always been more than a motor race. When Charles Faroux and Georges Durand dreamed up this race in the 1920s, they understood that watching cars cruise by for 24 hours, especially at night, could bore spectators, so they added a sideshow circus. Countless bars and beer tents served up German sausages, crepes, oysters, and French fries. Ham on French bread: 30 cents. Crowds lined up to ride the massive Ferris wheel that, lit brightly against the night, could be seen spinning incessantly from miles away. Guys ogled the packs of young French women, dressed fashionably in striped "poor boy" sweaters and tight pants belted below the waistline.

The sideshows lent the party an element of Kafkaesque absurdity. "In a tent," one man wandering that night described, "girl snake charmers charmed snakes that looked suspiciously stuffed while awed hundreds passed by at one franc a head. In another tent two girls lay in a glass box. They were said to be Siamese twins. It cost one franc a glimpse, and they did not even look related." A few yards away in another tent, strippers danced and grinded all through the night in a display of endurance that rivaled what was happening on the racetrack. Through it all came the cry of engines, and the faint smell of exhaust.

As the night wore on, the fields around the track became a camping ground. Tents offered some shelter. Others slept in the dirt with newspapers over their faces, or curled up in the backseats of cars.

By 1:00 A.M., twenty of the fifty-five cars had dropped out of the race. ABC's Jim McKay was still at it in the press box, stubble darkening his jaw line. "It's the middle of the night here," McKay barked into his microphone, "and the leader is the favored car, the factory Ferrari driven by John Surtees and his partner Lorenzo Bandini, who was one of the two winning drivers last year. That first-place car is followed by two more Ferraris. However, of very much interest is the fourth-place car, the #5 Cobra driven by Dan Gurney and Bob Bondurant of the United States. That car is not only in fourth place but is leading the GT division. And in fifth place, a remarkable story, is the one remaining Ford in this race, driven by Phil Hill and his partner Bruce McLaren from New Zealand. That car has moved up from forty-fourth place. It's going faster than any other car by far, lapping faster and faster every time . . ."

Death

At the kink near the White House, out of sight of the grandstands, the high-pitched wail of a Ferrari V12 clashed with the throatier bellow of a Cobra V8. The drivers were battling for position when the Cobra* blew a tire and clipped the Ferrari. Both drivers looked out their windshields and saw the world spinning. The shrill screech of burning rubber filled their ears. They wrestled with their cars, using all their tools—brake, clutch, steering wheel, gas. Sen-

* This Cobra was entered not by Shelby but by A. C. Cars of England.

tience reached its absolute peak and both men were suddenly suspended in time.

"A wonderful thing happens," Masten Gregory once said about losing control of a car. "Time slows down to a crawl, or else your mind runs like a computer, you know everything that's going on, and you can just sit there and consider the alternatives that will get you out of it." And when every attempt to regain control fails, there is always God. Bruce McLaren: "There's nothing like that blank flash of despair when it dawns on you that you might be going to hit something hard and there isn't a thing you can do about it. Except to get down in the cockpit and pray."

The Cobra flipped and tumbled off the road, landing upside down in thick bushes in an area forbidden to spectators. The Ferrari spun wildly in a cloud of smoke and ended up in the grass. Pieces of the cars littered the pavement and the nearest flagmen waved yellow in the night. Track stewards and medical staff were alerted. Miraculously, both drivers pulled themselves out of their cars with only minor injuries. A man looked at the Cobra and saw something under it in the thick brush. Was it . . . ? He looked closer.

There was a small body under the car.

A closer look: there was more than one body.

Police arrived along with reporters and medics. They found three young boys under the wrecked Cobra. The kids had sneaked under a nearby fence to get close to the track and they were watching under the cover of the bushes. None of them had any identification and all were pronounced dead.

Dawn

At 5:20 A.M., Phil Hill set a lap record. In the mist of a morning fog, he conquered the circuit in 3:49.2.* Minutes later he pulled the Ford into the pit with gearbox problems. The team of mechanics

* The lap record during the race was kept separate from the lap record during qualifying. Qualifying laps could be taken in flat-out sprints with less worry of overtaxing the engine or brakes. Thus Hill's race record of 3:49.2 was slower than Surtees's qualifying record of 3:42.

was exasperated, as was the crew of Ford executives. The Italian-made transmission was once again the culprit. Hill stepped out of the car, and as the early-dawn light illuminated his face, he stood there for a moment with his helmet in his hand. The sleepy crowd gave him a round of applause and he couldn't help but smile.

The race was barely more than half over, and the Ford team was finished.

Shortly after Hill's Ford retired, Surtees pulled his first-place 330 P into the pit. His car was limping also. He complained to the mechanics of a slipping clutch and the needle on the water temperature gauge was steadily rising. When the mechanics popped open the radiator cap, steam piped out. Surtees was exhausted and pissed off. First place slipped away. The technicians knew Ferrari would be angry, too; they'd hear it from him for sure when they got back to the factory.

"Why didn't we find this out in our tests? We run these engines on the dynos all day long!"

"Ah," they would reply, *"but there's something about going down that long straight at Le Mans that is different."*

By the time Surtees was in the car again, he was lying third. Ferraris held seven of the top eight places.

In fourth place, snarling along through the fog, was a Shelby American Cobra. In his pit, Shelby watched the cars roll by. The deeper into the race, the slower the hours seemed to pass. The crew signaled for Dan Gurney to bring the #5 Cobra in for repairs, fuel, and driver change. They were holding their breath. Gurney had slaughtered the GT lap record and was in first place in the GT class, but about an hour earlier, the car had started bleeding oil. The oil cooler had sprung a leak. Chief engineer Remington rigged a quick fix. Rules stated that a team could add oil only every twenty-five laps, so if the oil leak continued, the engine would seize and Shelby would have to pack it in. Gurney stepped out of the car and huddled with Shelby and driver Bob Bondurant.

"Brakes okay?" asked Bondurant.

"Yeah," Gurney said, "but I wouldn't trust 'em."

Shelby told Bondurant not to ride the engine too hard. "Watch your oil pressure," he said. He gave the driver a shove and Bondurant was off.

Finish

The final hours stretched out in a blur of speed, smoke, and noise. The crowds grew restless and the mercury in thermometers spiked. As the Dutray clock ticked past 3:45 P.M., the order of placement was all but set and the drivers slowed to assure their finish. The first-place car was five laps ahead of the second-place car, which was seven laps ahead of the third. Spectators not used to this tradition found it odd: at the end of the world's most brutal automobile race, the cars cruised slowly, no faster than they might on these public roads on any other day of the year. In the final minutes no driver would take the chance of blowing his engine or shredding a tire. Finishing was all that mattered now. The crowds leaned in, awaiting the moment when the checkered flag would wave and the champions would be crowned.

Just after 4:00 P.M., the red prototype Ferrari of Sicilian Nino Vaccarella and Frenchman Jean Guichet rolled over the finish line, winners of the 1964 24 Hours of Le Mans. Enzo Ferrari's cars finished in five of the top six places. Surtees placed third in a wounded 330 P. In fourth place, winning the GT class, was a Shelby Cobra. None of the Ford prototypes finished. Phil Hill, Bruce McLaren—they were no more than spectators at the finish.

Fans and media flooded the pavement, swarming the winning car. Ferrari flags waved high. The winners stepped toward the podium and soon the Italian national anthem was playing over the loudspeakers. The Shelby crew gathered around the Cobra, which had a California license plate on the lower right side of its rear end. Stirling Moss was there with a *Wide World of Sports* camera crew to interview the drivers, Gurney and Bondurant.

"Congratulations Bob," Moss said. "History I reckon has been made here today. How are you feeling, how much sleep did you get?"

"About five hours," Bondurant said.

"How about you?" Moss said to Gurney.

"About three, I think."

Up walked a skinny man in a distinctly Germanic hat. It was Huschke von Hanstein, competition manager of the Porsche team. He hugged Bondurant. "Thank you for beating them," he said.

The Americans weren't the only ones out to dethrone Enzo Ferrari.

A few yards away, Shelby stood, his curled bouffant looking a tad less than perfect. His team members crowded around, fists pumping toward the sky. Just two years had passed since Shelby first stepped foot in Iacocca's office with the idea of building his own car. Nobody believed he'd ever beat the Corvettes. Nobody believed his cars would finish the 24-hour grind at Le Mans. Now the "Powered by Ford" Shelby Cobra had placed fourth at Le Mans and first in the GT class. Shelby's Cobra was the Cassius Clay of motor racing, easy on the eyes and capable of the impossible. The reporters awaited comment from the Texan. He was always good for a quote.

"Fourth isn't bad," Shelby said. "Maybe America didn't hammer any nails in Ferrari's coffin this time. But we threw a scare into him. Next year we'll have his hide."

AFTERMATH

JUNE–DECEMBER 1964

ON MONDAY MORNING, the day after, the Ford team was summoned to the Hotel de Paris in downtown Le Mans. Tired men filed into the hotel smelling of grease and hangover. They found a Ford executive by the name of Leo Beebe waiting for them. He had a thin, bony face and the tall, gangly frame of a one-time high school basketball star. There was an air of sober intensity about him. He looked, in John Wyer's opinion, "like an evangelist missionary."

Most of the team recognized Beebe. He had stood mysteriously in the Ford pit in a suit and tie during the course of the race, arms folded, eyes taking it in, lips locked around a big cigar. If he had said a single word the whole 24 hours, few had heard it. The forty-six-year-old Ford executive was Henry Ford II's eyes and ears.

Beebe and Henry II had met during World War II. They'd served together in the Navy and had formed the kind of bond men do when in uniform during wartime. Beebe was no auto man. He was a high school sports coach. But Henry took him in and made him his top troubleshooter. Beebe had been with the company ever since, nearly twenty years now. "If he told me to jump out of that window, I'd do it, and think about it on the way down," Beebe said of his boss. He'd never seen an automobile race before the day he was made chief executive in charge of all Ford racing — stock cars, Indy cars, sports cars, the works. Henry II needed a man in charge he could trust.

At Le Mans, team manager Wyer believed the cars had performed well on their debut.

Leo Beebe didn't agree.

"I don't know anything about racing," Beebe prefaced his remarks. But you didn't need an engineering degree to know that Henry II's cars had lost. And it wasn't a matter of an easy fix. Sure, the transmissions had blown. "You could lay it to a gearbox," Beebe reasoned, "but if the gearbox didn't work how can we know anything else would work?"

Henry II had spent a lot of money on these cars, and Beebe wanted a checkered flag as soon as possible. The next major international sports car race was the 12 Hours of Reims in the heart of France's Champagne country. It was two weeks away—July Fourth weekend, no better time to wave the Stars and Stripes.

The racers were dumbfounded. That left the Ford Advanced Vehicles team no time to develop the car. They'd be throwing the car back into the heat of competition against the Ferraris before it was ready, before it had been properly tested and prepared. Besides, Reims was a brutal and dangerous race, run over 12 hours starting at the stroke of midnight on a road course with fast, undulating straights. For the racers, it didn't add up. This was a matter of a marketing executive making racing decisions and overruling the engineers. But the argument was for naught. The evangelist minister had spoken.

Two weeks later, at 11:59 P.M. on July 4, the pits at Reims were lit up under the stars, the grandstands filled with racing fans passing around chilled bottles of Moët. Wyer had entered three cars, with Phil Hill heading up the team. The blast of thirty-seven engines igniting at precisely midnight sent every creature in the surrounding countryside ducking for cover.

By sunrise, halfway through the race, the Ford team was heading back to their hotel in town. All three cars had suffered mechanical failure. The Ferraris raced on, taking first, second, third, and fourth places.

On a rainy September morning, Enzo Ferrari climbed into his Fiat and his chauffeur Pepino steered north up Highway E35, windshield wipers scraping at the glass. Through the Po Valley, the silhouettes of tractors could be seen trudging through the fog in the fields. They passed the old cities of Parma and Piacenza to Monza, a small city bordering Milan. The date was September 5, 1964, the day before the Italian Grand Prix. Each year the Grand Prix at Monza united Italians behind Ferrari from Turin to Palermo, and each year the Pope of the North made his appearance in the pit the day before.

The very word—Monza—was a synonym for speed. The 3.6-mile Autodrome was one of the oldest purpose-built closed racing circuits in the world, completed in 1922 and further expanded through the years. Cynics used another moniker to describe Monza—The Death Circuit. It had claimed so many over the years: Giaccone, Sivocci, Zborowski, Materassi, Arcangeli, Compari, Borzacchini, Czaykowski, Ascari, Von Trips. With few exceptions, Ferrari had known all of these men. His pilgrimage here each September was in part to honor them.

In 1964, controversy heightened the anticipation leading up to the Grand Prix. Surtees was number one at Ferrari, and number two was Lorenzo Bandini, a twenty-eight-year-old who was Monza's hometown hero. Bandini had gotten his start working in a garage just a few miles from the Autodrome, in Milan. Team manager Eugenio Dragoni was also from Milan. It was no secret that Dragoni favored Bandini. He decried Surtees while praising Bandini to the boss. *The Ferrari champion should be an Italian,* Dragoni argued, and Bandini was fast. As much as Italians adored Surtees, he was not one of them by blood.

Politics played a major role in determining the outcome of races. Cars were like athletes—no two the same, each responding differently to the elements. The man given the fastest car had an obvious advantage, and Dragoni sought to put Bandini in the better ride. As Surtees continued his campaign to become the first Grand Prix champion on two wheels and four, he found himself pitted against his teammate late in the season and criticized in the papers by his own team manager.

"Any right-thinking Italian should be able to see," Dragoni had told reporters, "that in Bandini, Italy has a true World Champion." Surtees would later come to believe that Ferrari himself was planting negative stories in the papers, stirring up the controversy.

By the time Ferrari arrived back in Modena from Monza that evening, in his Fiat with Pepino at the wheel, Surtees had won the pole.

When the phone rang at the factory postrace the next day, Dragoni's voice came through over the wire. Bandini had placed third and Surtees had won in record speed. At the Autodrome, mobs lifted their hometown hero onto their shoulders so Bandini floated atop the surging tide. Surtees marched toward the podium parting the crowds, his goggles up on his white helmet, his face stained with soot. He was so focused, he barely registered a smile, even as the crowd chanted *Il Grande John.*

Back in Dearborn.

"What does Ferrari have that we don't?" asked Leo Beebe.

"I can tell you in a word," John Wyer answered. "Ferrari. One man who knows his mind instead of a committee."

"General Motors is run by the committee system," Beebe said, "and they are fairly successful."

"Yes," Wyer snapped. "But how many races have they won?"

Wyer wasn't making any friends. He was sitting with Ford's Special Vehicles Committee in an office meeting room. The place smelled like antiseptic floor cleaner, what company employees called "that Ford smell." The committee included Roy Lunn and Don Frey, among others. As far as Wyer was concerned, *he* was the committee. The fifty-four-year-old had spent his entire life in racing in Europe. He was Mr. Aston Martin, a Le Mans champion team manager. If the suits would lay off, he'd turn their car into a winner.

Thus far Ford had entered GT40s in three races. Not a single one had reached a finish line. Three cars had been destroyed; two that crashed in test sessions and the one that had caught fire at Le

Mans. The only bright spot was Shelby's success with the Cobras in the GT class. Beebe wanted decisions made and he wanted them made in Dearborn, not in England, where Wyer was based.

The committee had come up with a radical idea. In its arsenal, Ford Motor Company had a 427-cubic-inch engine. The huge engine was dominating NASCAR, starting from its first appearance, a win in Tiny Lund's Ford at the 1963 Daytona 500, the first race after Henry II had pulled out of Detroit's Safety Resolution. In terms of displacement, the 427 was more than twice the size of the engine in the Ferrari Surtees had raced at Reims. Enzo Ferrari was surely building an even faster prototype Le Mans car for 1965, and Ford's 427 was "a way to solve the problem with a sledgehammer," as Lunn put it. In stock car trim the engine could produce almost 500 horsepower.

Wyer shook his head furiously. The idea was ludicrous, he said. The Ford Le Mans car didn't need a bigger engine; it needed durability. Besides, no one had ever shoved an engine that big in the back of such a lightweight sports car. Europe was the domain of small, sophisticated engines. It would never work.

"More power is always welcome," Wyer said, his voice tinged with exasperation, "but not at the expense of development and durability. I understand the 427 engine weighs 600 pounds. This would result in a car weighing 3,000 pounds. For practical purposes it would be a new car. We'd be putting back the clock exactly 12 months and running the risk of going to Le Mans again with a car that was untested and untried."

It was a daring idea. Putting all the 427's torque and power to the pavement in such a lightweight vehicle would require incredibly durable components. No braking system on earth could slow down 3,000 speeding pounds. There was fuel consumption to think about. And there was the transmission. If the transmission couldn't hold together under the stress of 350 horsepower, was an additional 100-plus-horsepower engine going to help?

If all these problems could be solved, however, Ford would have a 225-mph racing car.

Dinner that night at the Dearborn Country Club spilled into another office meeting the next day, during which the committee made its decisions. Roy Lunn was to return to Dearborn from England and set up a new shop to build an experimental next-generation Le Mans car. The Ford might have been a lightweight European sports racing car, but it was going to have a huge pushrod power plant, a big fat Detroit NASCAR engine. Wyer had his own marching orders. He returned to England and began to prepare two of the existing Fords for the Nassau Speed Week in the Bahamas in early December. It was a minor race, without serious contention from Ferrari. Surely the team could bring Henry II a victory.

Wyer's face looked especially grim as he stood on the wharf in Nassau three months later, watching as two cars were wheeled off a cargo ship. They were late. While loading onto a plane in London one car was shoved into a pallet, resulting in front-end damage. To get to the Bahamas, the flight made stops in Montreal, then New York, and en route to Miami, the pilot received a message that there was a bomb onboard. The plane made an unexpected stop in Savannah, Georgia.

If the trip was a fiasco, the event itself was worse. Both Phil Hill and Bruce McLaren suffered mechanical failures. In the pit, Leo Beebe stood watching under the beating Caribbean sun. He'd made the trip from Dearborn, and he'd seen enough. At the end of the affair he called a meeting in a local hotel. He stared down Wyer and the technicians.

Beebe began with a familiar refrain: "I don't know anything about racing." Then he added, "But there is one thing that has become increasingly apparent to me in the past few months. You don't either!"

He needed to make a change. When Beebe got back to Dearborn, he placed a call to Los Angeles. He took Ford's prototype Le Mans racing campaign away from Wyer and put it in the hands of Carroll Shelby.

When the first Ford GT40 arrived at the Shelby American shop, via air freight to Los Angeles International Airport, employees gathered outside to take a look. It was a cold, drizzly day in mid-December 1964. The car was not what anyone expected. It was banged up and filthy, hardly the historically expensive, state-of-the-art racing machine it was billed to be.

Shelby walked onto the wet pavement outside his shop and stared down at the car. Racing Cobras was one thing; they were his machines, with his name on them. This car had Henry II's name on it alone. Shelby's first race with the Ford would be the Daytona Continental 2,000 Kilometers. He knew with a glance at the Ford that he and his team were going to have to work around the clock. Daytona was eight weeks away. The whole racing world would be watching to see if Shelby's band of hot-rodders were the real thing.

Shelby never claimed to be a technical genius. His strategy was to surround himself with talent and to inspire his men to achieve beyond what they believed they could. He had his chief engineer Phil Remington, the most underrated technician anywhere in the world, Shelby would argue. Next he needed a project manager.

"You want 'em?" he said to Carroll Smith, an ex-driver and Goodyear tire engineer who worked at the Venice shop.

"What do you mean?" Smith answered.

"Do you want to be the team manager for our GT40s?"

"Shit!" Smith said. "Well yes I do, but I've already agreed to work for John Wyer."

Shelby said, "John has released you."

The most responsibility would land in the hands of the development driver, the engineer and pilot who would live with the car 24/7, on the track and in the shop. Shelby put his competition manager Ken Miles on the job.

Before the team did anything, they gave the car a complete steam cleaning. Then they craned out the engine and swapped in their own Shelby-prepared Ford 289. They gave the car a new coat of paint: Shelby American colors, guardsman blue with two white racing stripes down the nose, roof, and deck. The racing stripes looked sharp but they also had a function. They enabled engineers to better judge the behavior of the car in turns. Once the paint was dry, the team loaded the car onto a trailer and drove 70 miles inland.

At Riverside International Raceway, Ken Miles was waiting. A tall, taught-muscled Englishman, he was wearing a blue hooded sweatshirt, black wraparound shades, a filthy pair of chinos, white socks, and tennis shoes. He had a short crop of brown hair and a large nose that had been remodeled by a broken bottle during a tavern brawl in his teen years. He watched as the crew rolled the Ford off the trailer. They opened up the engine compartment. Miles leaned up against the right rear wheel with both hands so the grooves in the Goodyear dug into his palms. He gazed down at the Ford V8, its carburetors and its tangle of exhaust pipes. His son Peter, fourteen, stood next to him looking through an open door into the cockpit.

Miles was forty-six, older than most of the young bloods at Shelby American. His job was to take raw cars and turn them into racing machines. Many who knew Miles thought he was a sarcastic bastard. Carroll Shelby knew Miles was a genius.

On that December day, Miles slid on his helmet and goggles, climbed into the Ford, and took to the track. A handful of Shelby American employees, including new team manager Carroll Smith and chief engineer Phil Remington, watched as Miles threw the car around the circuit. The drive didn't last long. He pulled back into the pit and stepped out. Everyone waited to hear his impressions. He shook his head and said, "It's bloody awful."

At the end of the 1964 season, the whole sporting world turned its eyes to Mexico City. The Mexican Grand Prix was the final Formula One race of the year and three drivers had the title in their sights. All Surtees had to do was place second or better. Fought on bumpy pavement at seven thousand feet—the elevation wreaking havoc on his fuel-injection system—Surtees was in third when he entered his last lap. He passed his teammate Lorenzo Bandini and crossed the finish line in second, becoming the first man ever crowned World Champion on two wheels and four.

He had never shaken so many hands, he recalled. There was Prince Philip of England, and the president of Mexico, who presented him with a gold Longines watch. When Surtees arrived back in Italy, there was a formal banquet awaiting. Eventually the time came to celebrate the Italian way. They would sit down and eat.

Across from the gate of the Ferrari factory, a small inn called Il Cavallino stood on the Via Abetone. Ferrari owned the place. He'd opened it for his lunches, for there was little else in Maranello. One afternoon after the 1964 racing season, Ferrari arrived with his lieutenant Gozzi and Surtees. Ferrari desired an *aperitivo* and instantly his favorite drinks appeared: Formula 1, Formula 2, and Formula 3 cocktails, each one more watered down than the next.

"Formula 1 is for hard men only," Ferrari warned.

Surtees picked up a Formula 3. This sent the old man into hysterics.

Six female students were visiting the factory that day and Ferrari invited them to the table. His affection for women was no secret. A waiter popped the corks from two bottles of Tocai and then the dining room filled with the aroma of traditional Modenese fare.

At the table, Ferrari showered Surtees in affection. While praising the local delicacies, he slipped his fork into a bite and fed it to Surtees while the girls and Gozzi looked on. No one had ever seen Ferrari pull one of his drivers so close. A confidant once described the old man's temperament as "closed, like a walnut." And yet here he was, fawning over the English *pilota* publicly.

After lunch, Surtees and Ferrari left the group behind and strolled together into the racing shop. The driver climbed into a Formula One car and wrapped his fingers around the steering wheel. When he looked over his shoulder, Ferrari was standing there on the red tile floor looking down at him, smiling like a father who'd just given his son a new bicycle.

Surtees was the prince of a strange kingdom. Ferrari had created a Shakespearean world where intrigue was always brewing and men sometimes paid for mistakes with their lives. All season long the tension had nagged Surtees, poked at his explosive temper. The clashing with team manager Dragoni, the challenge of number two Bandini. Who knew what role Ferrari himself was playing in it all. As Surtees looked at Ferrari's smiling face that day in the racing department, he must've wondered what the old man was thinking. Ferrari wrote in his memoirs: "The facial expression, smile or frown or whatever it might be, is merely a form of defense and should be taken only as such." Was Ferrari's affection genuine? Was he in tune with Dragoni's machinations? Was he in fact behind them himself?

Sitting in the red car, Surtees turned his face forward and saw the steering wheel in his hands. Ahead of him lay 1965.

PART III

SPEED RISING

13

HENRY II, SHELBY, AND DAYTONA

JANUARY–FEBRUARY 1965

> Grand Prix racing has hundreds of men and girls of all ages who follow the cars and drivers everywhere and who worship openly at the shrine. Drivers see a romantic, reflected image of themselves in the eyes of these people. There is awe and the most naked kind of admiration there. As he settles into his cramped machine, revving the engine up and down, tense, eyes glued on the starters' flag, the crowd gulping with excitement—at such a moment a driver feels himself a god. What is danger next to that?
>
> —ROBERT DALEY, *The Cruel Sport*

BY 1965, AMERICA WAS fully in the throes of a speed revolution. The baby boomers had gotten their driver's licenses and the roads were thick with pimple-faced pilots, enjoying for the first time the seductive freedom of the highway.

Young men in the early 1960s had come to a realization: They didn't want to live their fathers' lives. All the keep-it-in-your-pants repression of the 1950s—follow the rules, never question—had planted a seed of desire in the new generation, a lust for adventure. It was this kernel that Hugh Hefner tapped into with *Playboy* magazine, that Albert Broccoli did with his 007 movies, and that Detroit was now exploiting in full force. Not just Ford with the Mustang, but John DeLorean and his new Pontiac GTO (named for the Ferrari GTO) and Plymouth's new Barracuda. It was the dawn of the muscle-car era. Speed was nothing but sex. To hammer the accelerator was to do it in the road. To indulge in risk was to be set free.

The avatar of manhood was the racing driver, the ultimate figure of bravado and virility. Heretofore a hero exclusively among the cult of motor-sport enthusiasts, this athlete's fame went mainstream. For the first time, races were being aired on television with some frequency. The revolution was being televised.* Spectacular crashes, the thrill of the chase—cameras mainlined speed into the nation's living rooms.

Fifty million went to see car races in 1964, more than double the number that saw pro baseball games, putting the sport second only to horse racing in spectator popularity. In Hollywood, four major motion pictures were in the works with top actors portraying death-defying racers—James Garner, James Caan, Tony Curtis, Steve McQueen. "The sudden outpouring of automotive features reflects a curious car-mania that has overtaken the film colony," the *New York Times* reported in 1965, noting the country's "rampant automania." In the Bonneville Salt Flats, Craig Breedlove was busting the land speed record in his jet-powered car Spirit of America, busting 600 mph in 1965.

All this served to tighten the strings that tied sport to industry. The iconography of racing drivers spurred the sale of fast cars, and fast cars heightened the popularity of racing. "Never before has a romance between man and machine blazed so strongly," wrote a *Los Angeles Times* columnist.

In Dearborn, Ford executives were riding the wave. Three years after Henry II had pulled out of the Safety Resolution, he could stand up at the stockholders meeting and make the announcement: "The company is now enjoying the most successful operations in its long history." He reported all-time-record sales ($9.67

* Readers may think the word "revolution" is over the top, but that word was used at the time, not just in terms of the emphasis on speed, but the way car companies were using market research for the first time to identify what car buyers wanted. See "The Detroit Revolution" in *Esquire*, July 1966.

billion worldwide over twelve months), all-time-record profit ($505.6 million worldwide), all-time-record employment (336,841 workers). The Mustang was on its way to becoming the best-selling car launch of all time. Iacocca already had his friend Shelby working on a racing version of the Mustang. ("Can you do it?" he'd asked Shelby. "I don't know," came the answer. "It's a secretary's car.") In newspapers all over the country, readers saw Ford ads with the company's Le Mans racing car ("The Ultimate Total Performance car") next to a Mustang ("The car it inspired").

On January 15, 1965, Henry II announced the promotion of Iacocca to vice president in charge of Ford, Lincoln, and Mercury cars and trucks. Iacocca packed his things and moved from Ford division headquarters on Michigan Avenue to the Glass House. The memory of his first job at Ford Motor Company was still fresh in Iacocca's memory. Now he was one of the most powerful men in Detroit.

Henry II installed Don Frey as top man of the Ford division. Frey never dreamed he'd make a paycheck of more than $100,000 a year. He'd always been the behind-the-scenes brain. Now reporters were coming to see *him*. He leaned back in his swivel chair and placed his feet on his desk next to a picture of his six kids.

"Six potential car buyers," he quipped. "It's a big, booming, glorious, red rosy market."

Henry II was making his own news. His lawyer issued a statement: "Mr. and Mrs. Henry Ford II have decided upon a legal separation." From that moment, the Italian mistress of America's most famous industrialist began to appear in newspapers and magazines seemingly for no reason, as no one would print that she was romantically linked to Henry II and his hundreds of millions. It wouldn't be proper, as they weren't yet married. No self-respecting journalist would print something that invaded a public figure's private life. Cristina Vettore Austin was photographed by Richard Avedon for a two-page spread in *Harper's Bazaar*, headlined "An Eclectic Beauty."

On February 19, 1965, Henry II married his Bambina at the

Shoreham Hotel in Washington. Immediately after, the couple caught Pan Am flight 100 to London. Their honeymoon, in Europe naturally, swept through St. Moritz to the French Riviera to Rome. Back in Dearborn, Henry II's old wife moved out of their Grosse Pointe mansion and his Italian wife moved in. Her exotic style, her very *foreign-ness,* made her the toast of Motor City instantly. To Henry II's chagrin, American cars were not her thing.

"Henry, look at that beautiful Mustang," she said one day.

"That is a Chevrolet," he answered. "I told you ten times, it's a Chevrolet."

Rebellion, sex, risk—Henry II's new persona embodied the spirit of the new Ford Motor Company. His wife symbolized his commitment to Europe and to change. One day in the not-so-distant future he'd take Cristina on another trip to Europe—to Le Mans, where, he believed, a Ford racing car (not a Chevrolet!) was going to make history.

Carroll Shelby dug his fingers into his thighs. He was sitting in the back of a small private plane with a couple of Ford executives. The engine buzzed like a gnat. At the helm, a Shelby engineer was piloting the plane onto a straightaway at Riverside International Raceway. The straight was a slightly downhill strip of pavement with a bridge at the end. As the bridge approached, the ground seemed to get farther away because of the gradient. The bridge kept getting closer and Shelby was getting nervous. He was himself a pilot, and he sensed they were headed for trouble.

Were they going to fly over the bridge or under it?

The pilot forced the plane down and it bounced back up into the air. By the time Shelby's feet touched pavement, he was ready to get on his knees and kiss the ground.

It was January 27, 1965, a Wednesday. Ford Motor Company was holding a press conference at Riverside to unveil the season's new machinery. Don Frey was there, as was Leo Beebe, and Shelby's presence always assured a good turnout. The gathering of reporters and photographers stood trackside as Beebe took to the podium and announced that Shelby American would be building and racing all of Ford's competition sports cars.

"We are taking this move to consolidate the construction and racing of all our GT-type vehicles within the same specialist organization," Beebe said.

Drivers paraded the new cars down a straightaway, cars that Ford would be racing and that, for the most part, customers could buy. First came the street version of the new Mustang GT350 built by Shelby ($4,547), hopped up with a Ford 289-cubic-inch engine, a four-barrel Holley carburetor, and a host of weight-reducing, performance-enhancing modifications. In all-out racing trim, the new Shelby Mustang went for $6,000. Then came the new Cobra with a 427-cubic-inch engine. Street version: $7,000, well more than a Jaguar XKE and nearly the cost of a Mercedes-Benz 230 SL. The racing Cobra cost as much as $9,000. For Ford-branded cars, these were incredibly expensive automobiles. Still, no one could imagine they'd fetch hundreds of thousands—if not millions—on the vintage car market in forty years time.

Finally, the crowd got a first look at the Ford GT40 in Shelby American colors, America's Ferrari fighter. It was not for sale, nor could anyone put an accurate figure on how much it had cost to build. For 1965, Beebe said, the GT40 would see action at a host of American and European races. The focus would be the big three: the Daytona Continental 2,000 Kilometers in February, the 12 Hours of Sebring in March, and Le Mans in June.

After the drive-by, the whole gaggle headed to the nearby Mission Inn for a schmooze-fest. The GT40s were trailered back to the City of Angels, to a new Shelby American plant, where Shelby's team was waiting to get their hands on it.

The new Shelby American factory made the old Venice shop look like a two-car garage. It sprawled over 12.5 acres bordering Los Angeles International Airport, consisting of two huge hangars where North American Aviation used to build Sabre military jets, a total of 96,000 square feet. The official move was March 1, 1965, but already the space was filling up fast. One hangar housed the racing shop and administration, the other the assembly line, where Cobras and Shelby Mustangs would be built by hand at a rate of roughly 125 cars a month. The pavement between these two hangars was already blackening with tire marks. All day the shriek of passenger jets taking off and landing at LAX rattled eardrums, and at night Shelby American test drivers gunned racing cars up and down the runways under the stars.

Three years after he had debuted his Cobra at the New York Auto Show, Shelby had become the largest independent sports car manufacturer in America, with sales edging up past $10 million a year. The business employed nearly two hundred people. Though Shelby had publicly declared Enzo Ferrari his nemesis, his organization bore a startling resemblance to the one in Maranello, Italy. Shelby had become the Ferrari of America, the charismatic man behind a small automobile company that produced handcrafted sports and racing cars. In contrast to Ferrari's reclusivity, however, Shelby was a bit of a showman.

He was on the road much of the time. When he was at his factory, he darted from task to task and office to office, his feet moving as fast as his mouth. As one visitor described him: "Shelby paces about restlessly like a lawman who expects trouble suddenly to bust out behind every swinging door in town." He moved so fast his secretary quit out of frustration. "You have to go 90 mph to keep up with him," she said, "and I'm just an old-fashioned 80-mph girl."

Now he's in his office on the phone: "Hello, butter bean. When I heard what you did, you could have cut buttonholes in my behind. My opinion may not be worth a pin whistle, but I think you're dumber than a hundred head of billy goats."

Now he's interviewing a new secretary: "How would you like to work in a snake pit for a real snake?"

So much change was afoot that spring. Some of the old guard from the Venice shop didn't take to the LAX facility. Men in suits worked there. Gone were the Friday drunken lunches at the Black Whale that lasted until closing time. Some of Shelby's employees were getting their notices from Uncle Sam. They were packing their bags not for the next race but for Vietnam.

Ken Miles was putting in the hours preparing GT40s to race Ferrari at the Daytona Continental. The team had eight weeks, and the cars had been taken apart so many times, they'd lost all their original design settings. "It may sound odd," Miles said, "but our first job was actually to get the cars back to where they had started."

One morning Miles and a crew from Shelby American headed to Willow Springs, a racetrack deep in the desert outside the town of Mojave, north of Los Angeles. The track had long straights and fast sweeping bends—just like Le Mans. It was a god-awful place. Everyone wore boots as it was rattlesnake country, and shades to protect the eyes from sand riding the wind. Technicians from Aeronutronic, a Ford-owned aerospace company, met the Shelby American team. The California-based outfit employed experts in the instrumentation and aerodynamics of missiles that traveled 18,000 mph. They were going to conduct an experiment.

The Aeronutronic technicians rigged a computer into the passenger seat of a GT40. It was space-man gear, the most sophisticated aeronautical equipment on earth, and it filled half the interior compartment. The computer sensors aimed to gather air pressure and temperature readings inside the car's ducting. The data would be transmitted to a trackside truck where a technician was stationed. An oscillograph would measure engine revolutions on paper right in the cockpit. It was almost certainly the first time computer equipment was used in the development of a racing car on the track.

Meanwhile Shelby's GT40 team manager Carroll Smith was going to gather similar information the old-fashioned way. Using Scotch tape, he stuck pieces of cotton yarn in tight rows all over the driver's side of the car (the right side). The movement of the yarn would tell the driver and anyone watching the direction of airflow when the car was in motion. Miles slipped on his helmet and lapped at slow speeds. A chase car followed, inside which a snapper aiming a Polaroid captured the movement of the yarn on the car's body.

According to the data from Willow Springs, the team discovered that at least 76 horsepower was being lost due to poor air ducting. Air was entering the car but it had no efficient exit. Over the next weeks, Miles and chief engineer Phil Remington redesigned the ducting and the lubricating system. They ditched the Italian wire-spoked wheels in favor of Halibrand magnesium wheels, shaving off thirty pounds, and fitted wider Goodyear tires on the rear, with lots of rubber that could grab at the pavement. They added larger front brakes so the car could be driven faster into turns. The engines were tuned to pump out 450 horsepower with gobs of torque. Though Shelby didn't know it at the time, some members of the team popped pills to keep the energy flowing. Amphetamines fueled all-night work sessions.

Reporters were showing up, eager to scoop the story. What was going on behind those closed doors? Unlike at Shelby's old Venice shop, where there was an open-door policy, everyone had to go through security to get on these premises.

"We have several advantages over other people who have played with the car," Miles told a reporter days before Daytona. "We can react to a suggestion—we can do something *right now.* We don't have to go through elaborate procedures of putting through formal design changes. If we decide we don't like something, we can take a hacksaw and cut it off. Practically everything we do is a panic operation. But if anyone can do it, we can."

Miles was accustomed to the grind. He'd spent the last two decades building cars and racing, moving from shop to shop and motel room to motel room. As a young man in England, he'd heard his calling in the howl of a sports car engine. He grabbed his wife Mollie and headed to Hollywood, a burgeoning hotbed of speed, and he'd been on the proverbial road ever since. During the 1950s, Miles became a cult figure in the Southern California sports car scene—amateur racing at its best. He was known for his delicate touch and lightning-fast reactions. He opened a small tuning shop next to a pizzeria on Vineland Avenue, just off the Hollywood Freeway. ("The same careful workmanship that has resulted in ten years of uninterrupted success in competition can make your car run better too.") But he was always short on money, and when the Internal Revenue Service padlocked his doors, he joined Shelby American in early 1963.

He lived with his wife and son in a little cottage on Sunday Trail, tucked into the Hollywood hills on rugged, winding roads off Mulholland Drive. His neighbors were accustomed to seeing his picture in the papers holding a trophy. They were used to seeing him jog by shirtless in the morning, the ropy muscles in his arms flexing with the weight of two-pound dumbbells. Out front of his cottage Miles usually had a Cobra or some other hot car parked, and when he drove his son to the hobby shop in town, the shop would empty out. A crowd would gather around the latest ride he was testing—hardly a Formula One car, but surely the hottest wheels in the neighborhood.

Miles had never gotten any offer to drive Formula One, never driven at Indianapolis. He'd never ridden that swaying train south through the Italian Riviera to meet with Enzo Ferrari in Modena. He'd given himself to a life of speed, and yet, strangely, the world of elite international racing had passed him by. As a technician and driver, he was for the most part self-taught. Now at forty-seven, for the first time in his life, he was working on a project with major financial backing.

There was a side to Miles that few of his colleagues, almost all of whom were younger than he, could understand. Miles was of that generation that had lived through World War II. He'd driven tanks in the British Army, on reconnaissance and recovery missions. His unit was among the first to pass through the Nazi death camp at Bergen-Belsen. His son Peter remembered his father telling the story of a time Miles ran around the corner of a building in the midst of combat right into a German officer. Both pulled their guns; Miles's lightning-fast reaction served him well. He'd lived through the kind of experiences that could shatter a man's nerves forever, or harden them until they became unbreakable. In the 1960s, Miles wore a filthy army jacket. He didn't talk much about the war but he wore that jacket like a flag on his back.

Working for Shelby, Miles wasn't supposed to race. He'd been hired as a competition manager, engineer, and test driver. But he couldn't stand it in the pit. He was building cars to beat Ferrari, and nobody was going to keep him out of the cockpit. He was going to compete.

One night Miles was working late in his office on the paperwork for entry at Le Mans. It was busywork, laborious and boring. Present was Shelby's photographer Dave Friedman, who had to take pictures of various auto parts to file with the papers. Talk of racing's death toll came up. Miles had seen his share of top-flight talents come and go.

"You know," he said, "I'd rather die in a racing car than get eaten up by cancer."

Miles had managed to stay alive over years battling on racetracks in cars he had built himself with virtually no safety equipment. His every breath was a testament to his skill.

On a February morning in 1965, four station wagons pulled into the empty infield of the Daytona International Speedway. Two transporters carrying four Cobras and two Ford GT40s followed. In the pit area doors wrenched open and out poured fourteen drivers, twelve mechanics, three engine technicians, two logistics managers, one team manager, one chief engineer, and Carroll Shelby himself in black pointed boots and a black cowboy hat that looked like it belonged to Jesse James. These men had haircuts and they were clean shaven. Shelby had given the orders. They might have been renegades, but now they were representing Henry Ford II.

They were first on the scene. Daytona was an American temple of speed. There was an eeriness about Daytona when its grandstands were empty. Miles pulled out a camera with a long telephoto lens that had a sticker on it—"Happiness Is a Hot Rod"—and gazed through the lens at the empty grandstands. Only a fraction of the thousands who'd soon fill those seats would know who he was.

Miles was on the verge, and he knew it.

The Daytona Continental 2,000 Kilometers was America's longest race, laid out over a 3.81-mile road course that winded through the oval track's infield. The race featured one thing that sports car drivers who didn't race stock or Indy cars weren't used to: the thirty-one-degree banked turns, so tall and steep they were a challenge to walk up. Driving through them was like driving across the side of a skyscraper.

Soon the pit lane was crowded with mechanics and the track filled with fast traffic. One by one, inexperienced drivers pulled their feet off their accelerators on that banked turn. Mortality was speaking to them. One Shelby American competitor pulled into the pit shaking with fear so bad he could barely step out of the car. The unmistakable song of an Italian V12 announced the arrival of World Champion John Surtees and the new Ferrari Le Mans prototype—the 330 P2, a 4.0-liter, dual-overhead-cam, midengine racer with a wing-like boom traversing an open cockpit.

Shelby approached Miles with another man in a cowboy hat. This, Shelby said, was to be Miles's teammate.

Miles eyed the stranger suspiciously. He had puppy-dog features, brown hair, and dull eyes. He was a late arrival, a Texan just like Shelby. When Shelby said his name, he pronounced it "Lawd Ree-oooby."

Miles had heard of Lloyd Ruby. He was an Indy-car driver, a Gasoline Alley veteran. Shelby could read Miles's mind. Miles had done round-the-clock development work on that GT40 and this cowpoke was going to screw up the car. Miles had no patience for anyone who didn't think as fast or drive as fast as he did.

Ruby headed over to the track hospital for his medical exam. When it was time, he ambled over to the GT40 wearing a helmet with a Texas Lone Star front and center like some kind of sheriff's badge. He moved so slowly, he seemed drunk. From the moment

Ruby took the Ford onto the track, however, he was flat out. When he flew into that banked turn, the Shelby crew stood and watched him.

"Jesus Christ," someone said.

Lloyd Ruby could drive.

On race-day morning, Shelby got his team together early. Leo Beebe had flown in from Dearborn. "This is a team effort," Shelby began, the drivers gathered around him. "The goal is to finish as many cars as high up as possible. Just let things take their natural course. If you happen to be in front, fine. If you happen to have an extra long pit stop that puts you back to fourth, I'll give you instructions as to whether you should try to pick up time or hold your position."

At 10:00 A.M., as thousands watched from the grandstands, three dozen cars thundered down the opening straight. Shelby stood in the infield, the brim of his cowboy hat shielding his face from the sun. He kept his eye on Surtees, who easily led the race. This track brutalized cars, and Shelby knew his only chance was if Surtees broke down. Then it happened: a tire blown on the banking. Surtees fishtailed in a cloud of smoke and dirt onto the grass. His Ferrari was done for the day. Flinging rubber had wrecked some of its bodywork.

Switching off, Miles and his teammate Ruby took command of the race. Miles knew how to maneuver that banked turn—foot down, feeling the cracks in the pavement jimmy the suspension. The sight of Miles jetting into the banking at 190 mph got the crowds on their feet. As the sun began to set, Shelby noticed something strange. When Miles pulled into the pit, he leaned in to talk to Ruby. Miles had an English accent so heavy, some couldn't understand a word he said. Ruby's Texas drawl was so thick, he needed an interpreter himself. Miles couldn't sit still. A nuke could detonate near Ruby and he might not blink his eyes. But they understood each other on a deeper level. Miles called the shots and Ruby had no problems. The car ran perfectly, smooth and fast.

It was after 10:00 P.M. when the checkered flag waved. Shelby's cars placed first through fifth. Eight weeks of preparation, and they had finally brought Henry II a checkered flag.

Down at Victory Lane, Miss Universe, Corinna Tsopei of Greece, presented Miles and Ruby their trophy. Then the party started. "I got drunker than shit," Shelby later remembered. Running on empty—no food and little sleep—it didn't take much. Miles wasn't a big drinker, but Ruby could crawl all the way into a Jack Daniel's bottle. Leo Beebe's cigar poked from out of an ear-to-ear smile.

The next morning, the front pages of newspapers across the country detailed the world news. President Johnson's precarious position in Vietnam was causing panic in the Capitol. Martin Luther King was calling for peace following the shocking assassination of Malcolm X in Harlem. In the sports pages, readers found something to smile about.

An American car had won a sanctioned international race for the first time in more than forty years.

220 MPH

FEBRUARY–JUNE 1965

At the Cavallino, a waiter slid a plate onto a table in front of Enzo Ferrari—a simple dish of fish and rice. Ferrari had invited a journalist from *The New Yorker* to sit with him at lunch. It was a rare occasion: he was in the mood to talk, and he unburdened himself on an array of philosophical topics.

"Women are more intelligent and dominating than men," Ferrari said. "Men are creatures of their passions, and this makes them victims of women. Ettore Bugatti, a great driver and racing car builder, and a fine gentleman, once told me, 'The perfect machine does not exist, mechanically speaking. The only perfect machine is a woman.'"

Did Ferrari have friends?

"There is something disgusting about the word *friendship*. To me it represents something sublime, but I have not found it so in reality. I have no friends."

A social life?

"No, none. Life passes soon enough. If you want to do one thing well, you have to work at it fast. A Ferrari may not be a masterpiece in exactly the same way that a great work of painting or sculpture is. It represents the work of many men bringing to life the ideas of Ferrari."

Religion?

"I prefer living in a state of problems and contradictions to believing in religion. You must be courageous in looking at the truth. Only he who looks at the truth is a whole man. He who says, 'Everything is good—let's go on,' is a fool. There are two kinds of people who do not ask themselves questions—those who are without conscience, and those who are religious. To me, religion is action. Sometimes, when driving on the *autostrada* at night, seeing the stars, I feel that there must be something infinitely grand that created this gigantic mechanism—but something that has no bodily existence. Not like what they teach you about religion in the schools. Moreover, I do not believe that this power, if it exists, is necessarily benign. It may also be evil."

In Modena, Surtees walked the streets greeting well-wishers and signing autographs. A sighting of the World Champion striding the cobbles brought great pride to the locals. Italy was a nation of hero worshippers, and Surtees embraced the fanfare. When he walked into cafés—an Englishman in Italy—people stood and applauded for him.

But as the 1965 season began to unfold, the tension at the Ferrari factory began to tighten once again. Surtees spent days developing the new Le Mans car, breathing in the rubber dust at the Modena Autodrome, working with the technicians at the factory. As the warm weather arrived, he and his wife moved from one exotic city to the next. At the Grand Prix of Monaco there were celebrities and the beautiful harbor full of yachts festooned with lights. At the French Grand Prix the top qualifier won one hundred bottles of champagne. But there was barely a second to soak in the glamour. Surtees would be steering into a bend at 150 mph at one moment, then it was off to the airport for the next town.

Disastrous starts plagued Surtees all spring long. Broken engines, bent metal. Memories of the 1964 World Championship season were fading fast. Team manager Dragoni seized the opportunity, attacking the English driver while trying to elevate his Italian Lorenzo Bandini to number one on the team. Surtees's relationship with Dragoni was like a tire worn down to the canvas, ready to blow, the consequences dangerous. The bitching match played out in the press. *"La polemica di Surtees favoriva Bandini?"*

The World Champion refused to accept blame for the Formula

One team's disappointing performance. It was this rivalry with the Americans, he argued. The F1 campaign was being neglected because Ferrari was allocating his resources to the sports cars, to fending off Ford and Shelby. "Naturally, from a commercial point of view, Ferrari has to consider this first," Surtees said bitterly in a filmed interview that could not have endeared him to his colleagues at the factory. "F1 comes last."

For Ferrari, it was the usual: interteam feuds, problems, and contradictions. There was conflict at every turn. As the next clash with the Americans approached, Ferrari took on the Fédération Internationale de l'Automobile (FIA), international racing's governing body, over a new GT called the 275 LM (for Le Mans). To race a car in the GT class, Ferrari had to prove he'd built one hundred examples (for this was a production car, a customer car). He hadn't built anywhere near one hundred. The FIA had let Ferrari slide in the past, but not this time. The 275 LM would have to race as a prototype, making it obsolete. Ferrari had spent many *lire* and work hours on the car. He played a trump card. He put out a statement and the headlines appeared all over Europe and North America.

"Ferrari Cars Quit World Title Meets."

What was racing without Ferrari? He was so powerful, he could pull the plug on the whole sport single-handedly. But he was bluffing and everyone knew it.

He had begun to take the Western invasion very seriously. Ferrari was stunned by the amazing amounts of money Henry II was spending on the Le Mans campaign in 1965. Beating the Americans was priority number one. The next duel with Ford was the 12 Hours of Sebring in Florida, where Ferraris had won the last four years in a row, six of the last seven. At the factory, the men were readying the cars for the journey across the Atlantic. But at the last minute, Ferrari heard rumors that displeased him. Something was up at Sebring, something he didn't appreciate at all.

The phone rang in Carroll Shelby's office. It was Alec Ulmann, the man behind the 12 Hours of Sebring in Florida.

"Do you mind if the Chaparral runs?" Ulmann asked.

Shelby held the phone to his ear and thought for a moment. The Chaparral was a fast roadster built by a Texan named Jim Hall. Shelby wasn't the only entrepreneur trying to build a racing sports car to beat the Ferraris. Hall was a serious talent and innovator, and his car was getting all kinds of press. The thing didn't meet FIA regulations as a prototype-class car by any means. It wasn't really street legal, nor did it have trunk space. It did, however, have a big Chevrolet engine. General Motors, the biggest car company in the world and an antiracing stalwart, was rumored to be funding Hall's project under the table. The buzz around Hall's car was as loud as that Chevy V8, and an appearance at Sebring would pit Ford against Chevy, and Shelby against Hall. Who cared if the Chaparral didn't meet regulations? If Ulmann could put Hall's car on the starting line at Sebring, he'd see a major boost at the ticket office.

"I been losing money," Ulmann told Shelby, "and I really need this."

Shelby was in a spot. He felt for Ulmann. And Jim Hall was the brother of Dick Hall, an oilman who once lent Shelby a bundle of money.

"I don't give a shit if the Chaparral runs," Shelby barked into the phone. "As long as you don't give Jim Hall the trophy and you tell the world that it is a 1,500-pound car and we have to weigh over 2,000 pounds." He couldn't be clearer: If Hall's Chaparral won, the world would know it wasn't an official prototype entry. Before he even hung up the phone, Shelby had a feeling he'd made a major blunder.

Sebring's population multiplied by six times the day of the race. It was a strange little town nestled among the swamps and orange groves of central Florida. At the 10:00 A.M. start, the temperature hit ninety-five degrees. Enzo Ferrari had pulled out of the race when he'd heard the Chaparral would be allowed to run, because he knew the car would win. Ferrari didn't want to be embarrassed by a car that was faster than his simply because it broke FIA rules. From the opening laps on Sebring's 5.2-mile track, laid out on a perfectly flat old military airfield, there was no catching the Chaparral.

At 3:25 P.M., halfway through the race, a tropical deluge pounded central Florida. In minutes the track was flooded. In some places water was six inches deep. Visibility shrunk to a few feet. Race fans ran for cover; they couldn't see the cars anyway. Shelby saw Phil Hill bring a Ford GT40 into the pit. When Hill opened his door, water poured out of the interior. Hill tried to get out but Shelby pushed him back in. A mechanic leaned in between Hill's legs and punched holes in the floorboards with a hammer to drain the water. "Watch out for my balls!" Hill screamed.

He motored back onto the track, where the pace had slowed to a crawl. There was a commotion. Shelby's truck driver Red Pierce was spotted unconscious, face down in the water in the paddock. He'd been electrocuted by a soaked generator.

The race was a disaster. By the time Shelby was in his hotel room taking a shower, four drivers, one truck driver, and two spectators were nursing wounds in a nearby hospital. Jim Hall's Chaparral won the race, snapping Ferrari's dynasty at Sebring, and though officially the prototype-class trophy was given to Ford racers Ken Miles and Bruce McLaren, the Chaparral stole national headlines.

"Hall Ushers in New Era at Sebring."

Shelby was furious. So was Leo Beebe.

After Sebring was the Le Mans test weekend. Ford executives stood in the pit and watched John Surtees obliterate the lap record in the new Ferrari 330 P2. Surtees looped the twisty 8.36 miles averaging 139.98 mph. Next came the Inter-Europa Cup 1,000 Kilometers at Monza on April 25. Then the Nürburgring 1,000 Kilometers in Germany on May 23. The Ferraris were winning everything.

As the spring days tumbled by, casualties continued to lure the ire of critics and bring controversy to the sport. Lloyd "Lucky" Casner at the Le Mans trials in a Maserati. Tommy Spychiger at Monza in a privateer Ferrari. Meanwhile, the publicity machine behind the upcoming Ford/Ferrari duel at Le Mans picked up speed.

One morning Shelby awoke in his bedroom in Playa del Rey

with his heart pounding in his ears. His angina was killing him. He'd had knee surgery not long before, and he was still nursing a bad wrist he'd mangled in a crash at the 1954 Carrera Panamericana. That wrist had required eight months of operations. Before the bones had fully set Shelby had raced again, battling at high speed with his broken arm taped to the wheel so he could steer. Eleven years later he could still feel the sting.

He lay there taking stock of his life. He was producing street and racing Mustangs, street and racing Cobras, and fielding a Cobra racing team that had a shot at beating Ferrari to win the GT class World Championship. His GT40s were about to make a historic run at Le Mans. He was a Goodyear distributor to the western states, moving $40,000 of tires monthly. He was launching All American Racers with Dan Gurney, the first-ever American Formula One team. Shelby was traveling constantly, and he drank a lot of liquor when he flew to help him sleep. In a single week in April leading up to the Le Mans test weekend, he flew from Los Angeles to France and back three times.

It was too much. Lately he'd been feeling out of breath a lot. Five years had passed since the doctors told Shelby he had a bad heart, that he might have no more than five years to live. He felt as though, at any moment, his heart was going to give out on him.

Don Frey's secretary buzzed him at his office at Ford division headquarters. Roy Lunn was on the horn, calling from his shop Kar Kraft, where he had been working on the next-generation Ford Le Mans car.

"I got something I'd like to show you," Lunn told Frey.

"Do you want me to come over there or can you bring it here?" Frey asked.

"I'd rather you came over. You've never seen what we're doing down here."

Frey arrived at Kar Kraft minutes later and had a look around. Lunn had stepped foot on the premises for the first time only a few months earlier, and Frey couldn't believe the operation was so fully outfitted already. All the equipment was spread out over the 4,000-square-foot space, even the surface table atop which designers build prototype cars—huge cast-iron slabs eight by fourteen feet, weighing thousands of pounds. Lunn had two of them. Frey didn't bother to ask where they'd come from. Truth was, Lunn had sweet-talked them out of the manufacturing guys. If he'd put in the paperwork, it would've taken years. He essentially stole the surface tables. The whole shop came together this way.

While Shelby American was racing the GT40s during the first four months of 1965, Lunn and his team—two design technicians, two draftsmen, and a secretary—were building the new prototype with the 427-cubic-inch engine. There in the shop Frey saw the car.

"It doesn't look much different," Frey said, staring at the thing through his professorial spectacles.

Lunn lifted the engine cover in the rear and Frey's eyes bore down into that mass of metal. A hint of awe registered on his face. Mounted behind the cockpit of this car that stood no higher than his belly button was a huge 427-cubic-inch V8, the biggest production car engine Ford had ever built. It required larger exhaust. Massive pipes snaked out of the eight combustion chambers, feeding into two smoke-spitting barrels that pointed rearward like cannons.

Standing over the car, Lunn explained the nuances. They'd moved the seating position forward to make room for the engine. There was a bulge in the cockpit between the two seats where the water pump now lived. The wider rear tires Shelby was using required a larger front end to hold the spare; Lunn's team had redesigned the nose so it was longer with a gentler slope, giving it a blade-like sharpness. As for the engine: high-compression cylinder heads, aluminum intake manifolds, a special high-rpm camshaft. The power plant had conquered stock car racing (it would win forty-eight out of fifty-five Grand Nationals that year). But was it capable of success on the world stage?

From the look on Frey's face, Lunn knew he was impressed. Up to this point, both believed they'd never have time to build and develop the car in time for Le Mans in 1965. They'd figured it could be raced maybe as a backup, as an experiment, with an eye toward victory in 1966. But suddenly the idea seemed plausible.

It was May—one month until the race.

Lunn needed someone to put some miles on the car as soon as possible. He recruited a local B-level sports car competitor and Ford dealership owner named Tom Payne, who drove the car over Ford's Dearborn test track for the first time on a rainy spring day, coursing through slow laps to see how it handled. He liked the car, so Lunn began making arrangements for a serious shakedown.

That day came a week later, at the 5-mile, high-speed test track in Romeo, Michigan. The spring sky was clean and bare. A 20-mph wind cut across the pavement, swirling the hair of a small group of men who surrounded a 40-inch-tall racing car. During the week, Ford engineers used the Romeo proving ground to test passenger vehicles, but on weekends it was free and Lunn had commandeered the banked oval. Ken Miles and Phil Remington had flown in from Los Angeles, and Tom Payne was on hand as well.

The Romeo test facility was situated an hour's drive north of Detroit amid peach and apple farms. There was always the danger in such rural environs that a deer or some other creature would wander out of the woods in front of a speeding car.* Miles circled the machine and chatted with Remington. Payne looked on, nervously. From the moment the engine's crankshaft turned its first revolution, the exhaust note roared. The car was far louder than the previous incarnation of Ford's Le Mans car. Miles sniffed the exhaust and listened, opening up his senses. With no trophies, checkered flag, or purse, he was going to be the first to risk the mighty shot into a banked turn at full speed in a car this light with an engine this large mounted behind the cockpit.

* In 2006, the author piloted an Aston Martin DB9 into a banked turn at 140 mph on this same Romeo oval when a coyote ran onto the track. A near miss.

Payne warmed her up, taking easy loops around the oval. Around 11:00 A.M., he began to put his weight on the pedal. On the banking he slowed and the car seemed to turn itself. He clocked a lap at 180 mph, faster than he'd ever gone in his life.

The crew broke for sandwiches, then Miles stepped into the car. He was far more sophisticated than Payne, and after some laps he went over his impressions with Remington. More laps, more adjustments. They worked on suspension setups—trial and error at 200 mph and beyond. Miles began to dive deeper into the turns, and the car spoke to him. He could sense its vast power, summoning more and more of it, holding the steering wheel with the tips of his fingers. When he rocketed down the straight past the crew the car's blue and white profile blurred like film footage played at high speed. Lunn looked at a stopwatch.

Miles had turned a lap at 201.5 mph, hitting 210 on the straights—four times the Michigan speed limit.

The car was frighteningly fast. They began to wrap the session but Payne stepped in. He wanted to test the Ford in its final trim. The crew watched him speed through his first lap. On the second, he hit 200 mph on the straight. When he brought the car in, he was visibly scared.

"I suppose you'd like to know what the dials were reading," he said.

Lunn waited for Payne's next sentence.

"Well," Payne said, "I didn't have the nerve to look down at them."

Later the crew gathered at a nearby hotel for a debriefing.

"What does everybody think?" Lunn asked.

Miles answered, "That's the car I want to drive at Le Mans this year."

The decision was made to prepare two cars with the 427-cubic-inch engine for the 24-hour classic, now four weeks away. The new car was dubbed Ford Mk II. The one tested at Romeo was immediately shipped to Shelby's plant for testing and development. Meanwhile, in Dearborn, Lunn and his team went to work building a second car.

Miles lapped the new Mk II at Riverside. On that track's twists, it quickly revealed a weakness, some loss in cornering ability due to the extraordinary weight of the engine. The car was less agile now. But on that long straightaway at Le Mans, the Ford would outpace anything with ease. If it could hold together. The days ticked by and the work at Shelby's LAX facility moved at a critical pace. In early June, one week before departure for Le Mans, Shelby received an important visitor.

The Deuce arrived one morning with two dozen men in expensive suits in tow—the entire Ford Motor Company Board of Directors. The Board was holding a meeting at Shelby American's new facility. Shelby was exhausted from the pre-Le Mans rush, his health suffering. Still, he put his charisma to work. His charm appealed to rich and poor alike. The two men had met before, and Henry II liked Shelby. What Shelby had done for Ford Motor Company could not be measured in dollars and cents.

The boss was in a good mood. A few days earlier, on Memorial Day 1965, "The Flying Scot" Jimmy Clark won a historic victory at the Indianapolis 500 in a Lotus-Ford. Clark led 190 of the 200 laps, topping A. J. Foyt's record with an average of 150.686 mph. Henry II was swimming in great publicity. Ford had conquered NASCAR, and now Indy. There was but one race left on Henry II's agenda, and he had placed the responsibility of winning that race in Shelby's hands.

Shelby showed Mr. Ford around. Here were Mustang GT350s, racing Cobras, the two new Mk II Le Mans cars. Shelby was the kid in the candy shop. A Ford executive tapped on Henry II's shoulder and whispered in his ear, "Can you imagine a guy getting paid for doing this?" Henry II had spent his life in the auto business, making American cars. But he'd never seen American cars like these. Ken Miles took a break from work to take Henry II on a joy ride in a Cobra. The two fired around the premises, Miles letting the car have it, nearly taking out a fence in the process. It was a Board of Directors meeting Henry II would not forget.

On May 25, 1965, the Ferrari team gathered at Monza at dawn to run a final 24-hour shakedown of the Le Mans prototype. This test session aimed to replicate all the conditions of the race. The car would lap the track, stopping only for refueling, driver change, and maintenance. Pit technicians and drivers would perfect their fusion. Anything that could go wrong outside of mechanical failure, they would be ready for it.

Mechanics in jumpsuits wheeled a 330 P2 off a transporter and sparked its engine. The 4.0-liter V12 coughed and spit, then fired on all cylinders. (In American terms, the Ferrari engine displaced 244 cubic inches compared to Ford's 427.) In the pit, the tired-eyed mechanics unloaded their gear and smoked cigarettes. The weather was clear, but rain was on the way. At precisely 8:00 A.M., the first driver stepped into the Ferrari and took off onto the track.

Ferrari arrived at Monza later in the day. He distanced himself from the pit box, not wanting to interrupt the focus of his men, preferring to stand back with a friend, chatting and watching. The 330 P2 lapped over and over, Surtees setting the fastest time as usual. In the pit, Ferrari's men were calm. They almost seemed bored.

That afternoon a driver named Bruno Deserti approached Ferrari. Deserti was an up-and-comer recruited by Dragoni. He was local, from the Emilia. An *Italian.* His invitation to these trials was major news, landing his picture in the papers. ("Deserti is really fresh and young," Dragoni had said. "He has the will of iron, not to mention a slender physique and a notable resilience. He's tough.") Deserti wanted to thank Ferrari for the invitation, and he nervously showed the old man a new white helmet. He explained that he had purchased the helmet for the occasion as a good luck charm.

"It represents the crowning of my dream," he said.

Ferrari looked into the driver's brown eyes. Both knew this was the twenty-three-year-old's chance, that his dream of becoming a Ferrari racer—a dream countless boys and young men in Italy had in common—hinged on what he was about to do when he stepped into the cockpit. Ferrari could see Deserti was enslaved by his nerves. He'd been in the young man's shoes. He could remember trying out for Alfa Romeo so many years before, and he would never forget the purity of that emotion. Ferrari later described what he saw in Deserti's eyes that day at Monza: "When such passion connected to the human spirit explodes like this, it is stronger than life itself, and it is stronger than death." Ferrari smiled, said hello, then turned away. He feared that any further discussion would distract Deserti, so he appeared busy.

The test session rolled along. At 6:55 P.M., the car was in the pit for refueling: 140 liters of gasoline. Deserti felt the tap on his shoulder. He stood and slipped on his new white helmet. He stepped into the car, checked the brakes. And then he went. Ferrari watched him drive by as he began to lap at comfortable speeds, building his lap time: 1:58, 1:52, 1:50, 1:48, 1:47, 1:46. (Surtees lapped at 1:32 that afternoon.) Suddenly there was silence on the track. In the pit, driver Lorenzo Bandini was the only one who caught sight, out of the corner of his eye, of the red car rocketing off the track into the woods. He jumped to his feet.

"He's out!" Bandini screamed. "Let's go! Let's go!"

When the mechanics and drivers reached the car, they found it 50 meters into the woods, turned on its side. It was in tatters and smoking. Inside Deserti was slumped in the cockpit. When Ferrari arrived on the scene he looked down and saw the black stripes on the pavement leading into the upturned grass toward the woods, where the crumpled car remained. Those stripes represented the last-ditch effort of a young man clawing for his life.

Deserti died a few minutes after 7:00 P.M. By 8:00 P.M., Ferrari was in his Fiat, Pepino steering south for Modena. It was a long, quiet ride.

The Le Mans shakedown was over. For the team, the accident was devastating. The work was undone, their car unproven. Would the brakes last 24 hours? The suspension, gearbox, and engine? The car was *fast,* but there was an element of mystery to it, one that would come to haunt Ferrari when the flag dropped at Le Mans.

By early June, the world press was all over round two of the Ford/Ferrari war at Le Mans. "Enzo Ferrari believes that victory at Le Mans sells more of his cars than all other victories combined," the *New York Times* stated. The papers were calling Ford's quest to end Ferrari's dynasty the biggest American invasion of Europe since Normandy. "With the great increase in power in the last 12 months," wrote a columnist in England's *Autosport* magazine, "it is more than likely that the Ford and Ferrari prototypes will be capable of achieving 220 mph on the Mulsanne Straight. The sobering thought is that the men who will drive these projectiles must be carefully chosen, for it is certain that not even a Grand Prix driver has driven anything before which could attain this type of speed."

At Ferrari's factory, the big transporters with racing cars stacked on their reinforced spines lumbered out of the gate. The news of Ford's new big-engined Mk II had reached Maranello. "The pistons are as big as wine bottles," Ferrari was heard saying. At LAX, airport employees carefully wheeled Shelby American racing cars onto TWA big birds as Shelby and Leo Beebe looked on. The two posed for a ceremonial photograph. Shelby flashed his famous smile. Beebe had the look of a man standing under a noose.

Six Fords were entered (plus five Cobras) and ten Ferraris. The prognosticators had picked their winners. But no one was prepared for what they were about to see that weekend.

LE MANS, 1965

> Once, in my racing days, I was in third position when I suddenly saw a car ablaze on the edge of the track. I could make out the number: it was the car that had been just in front of me. What thoughts do you think passed through my head in that instant? Well, my first thought was: One less, now I am second. My second thought was: I wonder if he's hurt. And my third thought was: It could have been me.
>
> —Enzo Ferrari

On the night of Wednesday, June 16, 1965, World Champion John Surtees sat in a restaurant in Le Mans staring out the window. A violent storm had leveled trees and brought down power lines all over western France. For the first time in anyone's memory the opening day of practice on the circuit was canceled. Already things weren't going as planned.

Surtees quietly ate his dinner. The wait was a drag. He wanted to get in the car and get going. Someone asked him about last year's Ferrari Le Mans sweep.

"The car that won wasn't going the fastest," Surtees said. After all, he was fastest, and he didn't win. "But it kept going while the others had silly troubles." A gust swept through so fierce it threatened to shatter the window next to him. "I'll have to get out my slide rule to check on the flexibility of glass," he said.

Surtees had a lot on his mind. The secret was out; it had hit the papers, this "mid-season rift" between him and his team manager

Dragoni. And the 330 P2. It was lightning fast, but Surtees believed that the car had not been properly developed for a 24-hour competition. He knew the Americans were coming at him with some heavy machinery, and Ford's Le Mans lineup was world class: Phil Hill, Bruce McLaren, Dan Gurney. Surtees had battled these athletes in Formula One and knew how skilled they were, how hard they fought. All this, not to mention the mental preparation for the physical punishment of a 24-hour grind, the stress on the brain, the body, the nervous system.

How could any man endure such anticipation? Outside the rattling window, the rain continued to fall.

When Surtees finally stepped into the Ferrari pit the next morning, he was completely focused. He was wearing clean white coveralls and racing shoes, holding his helmet in his hand. Journalists and some die-hard fans attending the first day of practice watched the weekend's star step into his car. Though the track was still wet, it had been cleared of downed tree branches. The sun had appeared and the red Ferrari shined bright, the number nineteen painted on the nose and doors. Surtees's wife sat quietly, holding a stopwatch and clipboard. Dragoni stood hunched, saying little. The team manager wore thick-frame glasses over a striking nose; he looked like an older Peter Sellers. Surtees checked his instruments and revved the V12 engine. Its song brought such joy to enthusiasts, it was available on LP record. Tires dug in and Surtees was gone.

On the track he chose the perfect lines, owning every bend and corner, every gear shift and touch of the pedal. No man had ever made it around this 8.36-mile strip of pavement faster than Surtees. When he posted his best lap time that day, applause came from the Ferrari pit. He stepped from his car and the mechanics congratulated him.

"Bravo, bravo."

Surtees set a new mark on the first day of practice. Just three years earlier, drivers were challenged to be the first to break a four-minute lap here. Surtees's wife logged the time on her clipboard: 3:38.8, an average of 137.701 mph.

In the Ford pit, confusion reigned. Some fifty technicians were standing there, most wondering what it was they were supposed to be doing. Even the lineup of cars was confusing: two Ford Mk IIs with their 427 engines entered by Shelby American; five Cobras entered by four different teams; four GT40s with 289 engines entered by four European teams, one from Switzerland (though the car was prepared by Shelby American), one from Britain (prepared by Shelby American), one from John Wyer and Ford Advanced Vehicles, and one from Ford of France. So much for national colors; the cars were painted different colors so that each could be identified easily in sunlight and darkness, the two new Mk IIs in white and black. No fewer than 840 Goodyear tires had been shipped in. Once the first practice session got under way, the pull and tug between Shelby's men and the Dearborn engineers frayed everyone's nerves. Who was in charge?

Shelby's men don't have college degrees, the Dearborn suits argued. *They aren't real engineers.*

Phil Hill arrived in the pit that morning in the strange position of knowing very little about his car. He would be in one of the new big-engined Mk IIs. Ken Miles had performed last-minute development work, and though Miles was a brilliant engineer, there was no way he could've made the car bulletproof in that short amount of time. Hill looked over the vehicle and its heavy 427 before heading out onto the track.

It'd been an awful spring for Hill. His career was now in a complete state of collapse. He'd been known in previous seasons as one of the safest drivers, having never crashed seriously. But in the past couple months, competing in Europe for the Cooper Formula One team, Hill had pulled himself out of two smoldering wrecks and had blown a handful of very expensive engines. His boss John Cooper had cursed him out in public and had fired him. "There comes a time when every race driver becomes emotionally unsuited to this type of driving," Cooper had said. "Hill has reached this point. There may be some kind of driving Hill still can do, but I don't know what it is."

Just before heading off to Le Mans, Hill had read a profile of himself in the *Saturday Evening Post*. It showed an image of his soot-stained face next to a headline that read "Sundown of a Champion." The story told of "defeats and humiliation." It dragged Hill through his 1961 championship Grand Prix again, the last race of Wolfgang Von Trips. Hill hadn't so much come into his World Champion status as he had collided with it. Ever since, he seemed to recoil, falling deeper into a world of failure and danger.

That first morning of practice, Hill pulled the Ford Mk II into the pit and stepped out of the cockpit shaking his head. "It's absolutely frightening," he said. The 427 provided ungodly amounts of torque and power, he reported, but there was no stability. "If we could get it more stable, it'll go like a bat out of hell."

Roy Lunn had an idea. He turned to one of Shelby's men, Bill Eaton, and told him what he wanted. Eaton went behind the pits and got his hands on some sheet aluminum. He cut the sheets and bent them over the side of a trailer, fashioning two dorsal fins. He then riveted the fins to the Mk II's rear deck like the vertical stabilizers on a fighter jet. Soon there were more fins: eyebrows over the front wheel wells to keep the front end from taking flight. Hill motored back onto the track.

By Friday evening, the last practice session, the Mk II began to behave. Ken Miles and Bruce McLaren had posted competitive lap times. As the sun was setting, Hill pulled on his helmet. He had one last shot at a strong qualifying time. Shelby approached him. The two men had known each other for a long time. Shelby trusted Hill.

"Let it out," he said.

"All the way out?" Hill asked.

"All the way out."

Hill climbed into the car. Tucked behind him, inches from his spine, was a mechanism with the power of a herd of 450 horses. Out on the track he let those thoroughbreds go. In the pit, stopwatches clicked away the seconds. Necks craned to watch as Hill dashed down the pit straight. Unlike the Ferrari's violent high-pitched wail, the Ford 427's roar was throaty and thunderous. When Hill posted his best time, morale soared. Everyone in the pit suddenly realized that their car was faster than the Ferraris.

Handshakes and high fives. Phil Hill had won the pole at Le Mans.

The next morning's *New York Times:* "Phil Hill of Santa Monica, Calif., at the wheel of a 7-liter [427-cubic-inch] Ford prototype, shattered the lap record for the Le Mans track today when he toured the eight-mile circuit in 3 minutes 33 seconds at an average speed of 141.362 mph. Hill gave a great lift to the Ford cars, now favored to beat the Italian Ferraris in . . . the most grueling auto test in the world."

In the early afternoon on Saturday, June 19, ABC's Jim McKay stood on camera giving prerace commentary. Behind him the grandstands were loaded. Once again, the race was drawing unprecedented numbers, more than 300,000. The camera crew prepared to cue McKay. For the first time in history, a European sporting event was being piped into America's living rooms via satellite.

It wasn't taped. It was live.

McKay: "Why did we pick this event here for the first one to be brought to you live via early bird satellite? First of all it's the biggest sports event in Europe. Secondly, along with Indianapolis, it's the world's most famous automobile race."

Never had the French, English, and Germans heard so many American accents. Fans had crossed the Atlantic in droves. To them, guys like Shelby, Gurney, and Hill were true American heroes.

Behind the pits, Shelby was taking a breather with an old friend—Masten Gregory, "The Kansas City Flash." The two had known each other since the early 1950s. Gregory was from Missouri, Shelby from Texas, and in the early days of the postwar sports car Renaissance in America, they won everything between the two coasts, always in the latest machinery from Europe. At Le Mans, Gregory was going to race a year-old Ferrari 275 LM for Luigi Chinetti's North American Racing Team. Sure, it was a Ferrari, but it was a station wagon compared to the 330 P2 Surtees had. Gregory's car had no chance of winning. Gregory's eyes met Shelby's through his trademark Coke-bottle glasses.

Jesus Christ, he said in his Missouri drawl. Ah must say, this is some scene.

Shelby sure knew it. The old days—when a man showed up, tested his car, played some gin rummy, then raced—were over.

As the two stood talking shop, Gregory worked a cigarette between his fingers. Shelby loved ol' Masten. He hadn't changed a bit. His story was legendary in motor-sport circles. Masten Gregory had inherited a fortune at nineteen from his deceased father's insurance company. Already by that time he'd honed his skills drag racing on the streets of Kansas City in a hopped-up 1949 Ford (thus the "Kansas City Flash" nickname). He had moved to Rome with his wife and four kids in 1954, one of the first Americans ever to try to make it in Europe. He became famous as much for his aggressive style as for his bizarre crash technique. Twice he'd stepped out of cars speeding toward immovable objects. Simply stood up and jumped. Injured—very much so. But he'd survived. Most believed Masten Gregory was simply out of his mind. And those glasses! Without them, he couldn't see the steering wheel in front of him.

Now, in 1965, Gregory had long since blown his money on fast cars. He was thirty-three, and by his own admittance, he never believed he'd live past thirty. "I hadn't made any plans," he'd said. "I didn't think it was worthwhile." Few men appeared less competitive than Masten Gregory, until you looked deep into his eyes. And then it hit you. He had a monkey on his back. *Going as fast as a car could possibly go.* That's all that mattered. At Le Mans he was about to compete in a back-up car for the back-up Ferrari team. Looking at him, Shelby saw what he himself might've become if health problems hadn't forced him to retire: an aging talent riding the inertia of a once-great career.

They shook hands and parted ways. They were, after all, competing against each other. Just like the old days.

A few minutes before the start, an announcement came over the loudspeakers in French: "Drivers take your positions please." Fifty-one men lined up across from their cars at the base of the grandstands. Phil Hill was in the press box giving television commentary; his teammate Chris Amon, of New Zealand, would start the race. Ken Miles ambled restlessly in the Ford pit; he would take second shift after Bruce McLaren. Surtees stood ready to sprint across the road to his car. The sun was out; speed conditions were perfect. France's Sports Minister Maurice Herzog stood in the middle of the road holding the flag high. Television cameras were rolling, ushering in the era of instantaneous global electronic coverage of world events.

At 4:00 P.M. precisely, Herzog dropped the flag. The race began.

From the grandstands the fans watched the drivers sprint across the narrow two-lane road. The smoke, the noise, vehicles accelerating violently—no one could witness this spectacle and not think of war. Then, in a matter of seconds, the cars had disappeared under the Dunlop Bridge and into the Esses, leaving only the sound of the crowd's murmur and the commentator's shrill voice piping from the loudspeakers.

Some four minutes passed before the first cars came back into view. The crowds rose to their feet and leaned in, waiting to see who would loop around at the front of the pack at the end of lap one. The leaders appeared to be traveling slowly in the distance, two white and black cars followed by a red one. The engines could barely be heard. But as the cars approached the grandstands, the noise resounded and eyes adjusted to the relativity of speed.

It was one, two. Chris Amon and Bruce McLaren ran nose to tail in the two Shelby American Ford Mk IIs at roughly 195 mph. Surtees's Ferrari chased. The rest of the field followed—Ferraris, Cobras, Fords, Porsches, Triumphs, Alfa Romeos, an experimental turbine-engined British Rover . . .

Shelby settled in, shaking the early jitters loose. Out of the corner of his eye he saw Leo Beebe standing quietly. The early minutes had unfolded exactly according to plan. McLaren took front position and left Surtees in the dust. Lap after lap, the Fords added to their lead. News piped from the loudspeakers: McLaren had set a new lap record. A few minutes later, he lowered the mark yet again. After thirty-eight minutes, McLaren stretched his lead to thirty-eight seconds.

The crowds standing on either side of the track along the Esses watched the leader maneuver his Ford through, exhaust pipes firing showers of sparks as McLaren downshifted. Through that blistering downhill run into the uphill lefthander, the car hugged the pavement on the edge of adhesion. Every revolution of the camshaft, the engine gulped a great breath of air and exhaled the smoke. Spectators who saw that leading Ford power down the road could not help but wonder what it would feel like inside the cockpit. As Mark Twain once wrote of steamboat racing, "This is sport that makes a body's very liver curl."

But was McLaren moving too fast? In the press box Phil Hill watched and questioned.

"What do you think, Phil?" Jim McKay asked.

"This is quite a pace for so early in the race," Hill said ominously. "It's a bit quick."

A few minutes past 5:00 P.M., McLaren and Amon came in for scheduled pit stops, the lead over Surtees at fifty seconds. Ford's 427 engines guzzled fuel at five to six miles per gallon, well more than a gallon per lap, but the car's sheer speed would make up for time lost refueling. Shelby watched both drivers step out of the cockpits. They were laughing; they couldn't believe what was happening. Both claimed to stay 400 revs below max—which was 6,000 rpm and 6,500 on straights in top gear—as instructed. And yet they were murdering the Ferraris. Amon's take on the car: "It's like a rocket ship!"

Shelby spotted the first sign of an impending disaster in the third hour. As cars flew past, one Ford motored through the grandstands streaming white-hot smoke from its engine compartment.

Before the sun at Le Mans had set, and the challenge of night racing confronted the pilots and pit crews, a strange fever began to spread among the Ford engines. The contagion struck the GT40 driven by Bob Bondurant first. When he pulled his car into the pit, the mechanics could immediately smell the overheating metal. Exhaust pipes glowed red. The driver explained that his temperature gauge had hit 266 degrees Fahrenheit. Minutes later, another GT40 pulled in, the engine also overheating. And then a third. All the Fords with the smaller 289 engines were suddenly infected. No one knew the source of the fever, nor how to treat it. One by one, the Fords were wheeled out of the pits into the paddock, and an overwhelming sense of embarrassment set in.

At 8:00 P.M., McLaren pulled the leading Mk II into the pit to hand over to Ken Miles, who'd been itching to get into the race. Miles started with a comfortable lead, but as he ripped through his first laps, he began to feel his transmission choking. He was losing gears. All those late-night hours of work, the dangerous test laps turned at Romeo and Riverside—all of it for naught. Miles could've put his fist through the windshield. He was done for the weekend. His teammate McLaren was in the Shelby American tent eating dinner when he heard the news: The gearbox in his car had failed. McLaren would be home in London in time to watch the finish on television.

Four hours into the race, Ferraris were lying one through four. The Surtees car was running beautifully in first place. Shelby stood helplessly as the sun dropped toward the western horizon, his stomach souring. Already Ford mechanics and drivers were cataloging their equipment, packing it up, and heading back to the hotels. Beebe was avoiding eye contact.

"I can smell the chicken shit now," Shelby muttered to himself.

At sunset, Phil Hill was on the course trying to catch up to the

Ferraris in the only remaining Ford. His car had suffered some clutch problems, and he'd lost time. He was miles behind, and he knew his Mk II wasn't going to last. He could feel the clutch slipping. There was nothing left to do but floor it and try to squeak out a lap record, some piece of good news the team could walk away with. When he passed the start/finish line, he summoned all his expertise and all the power of the big 427 engine. Headlights were on now. He piloted through the Esses into the tight *Tertre Rouge* corner. On the Mulsanne Straight he was clocked at 218 mph. When he passed the start/finish again, the official timekeepers recorded the fastest race lap in Le Mans history, an average of 138.44 mph. But the news registered only a hint of pleasure on the faces of Ford staff. Soon Hill was guiding his car back to the pit, his clutch—and his race—destroyed.

It was barely 11:00 P.M. and all the Fords had broken down. The most anticipated automobile endurance race of the decade had fizzled not long after its start. All the posturing and all the company money spent had resulted in a gut-wrenching failure before hundreds of thousands and live television cameras.

The debacle left each Ford man strategizing to cover his ass. Someone was going to get fired. "You should have seen all of the finger pointing among the Ford people once things started unraveling," one team driver later said. "It was almost humorous." Shelby had nowhere to hide. He stood alone and defeated.

All over the grounds, the crowds' attention steered away from the action on the track. The party raged on. Drinks went down warm; ice had no chance in this heat. Under the cover of darkness, the race took another unexpected turn. Surtees was moving along at a quick clip when he felt his front suspension suddenly collapse. He felt the car buckle. He pulled his foot off the pedal and gently maneuvered back to his pit as the competition roared past him. Surtees was beginning to loathe this race. Silly little things, every time. Then, within minutes, more Ferraris pulled into the pits with shattered brake discs. The brakes were not surviving the abuse of the

Mulsanne Corner. Beneath the stars the mechanics jumped to action, piecing together disc brake systems with whatever parts they could find.

Surtees had foreseen these problems. None of this would've happened if it were not for Bruno Deserti's fatal accident. The frontline Ferraris began to fall further and further back, leaving the race wide open.

As spectators drifted off to sleep in the early hours of morning, one thirty-three-year-old American driver began to inch his way toward the lead in a tired and bruised Ferrari, entered by Luigi Chinetti's North American Racing Team. The pilot moved down the Mulsanne Straight with all his weight on the pedal, straining the car's every bolt and weld. The Kansas City Flash was lapping at a furious pace, driving flat out as always.

LE MANS, 1965

THE FINISH AND THE FALLOUT

LUIGI CHINETTI STOOD in the pits at dawn, silent and imperious, his dark brown eyes fixed on the rolling thunder. He had a widow's peak of gray hair, and 1950s-style spectacles. Many of the fans sitting nearby looked down at the compactly built sixty-four-year-old and had no idea who he was. But those who knew racing knew that Chinetti was as much a part of Le Mans heritage as the seats that cupped them. He was a three-time champion here (1932, 1934, 1949), a sly and beloved man in the automobile business, the little giant behind Ferrari in America. An Italian who'd defected to the United States during World War II, Chinetti had dedicated his life to the *idea of a car,* one that bore another man's name.

As head of Ferrari's North American Racing Team, Chinetti had recruited young talent that became legends in their time. He'd sold Phil Hill his first Ferrari. He'd given Dan Gurney his first Le Mans ride. He'd discovered the Rodriguez brothers from Mexico, Pedro (twenty at the time) and Ricardo (eighteen), the latter who perished at the Mexican Grand Prix in 1962 in front of his home crowd. Chinetti still mourned for him. He ran his team with a flare of old-school Italian style. "Racing with Chinetti was an experience in itself," one driver once commented. "He never told you anything. I was afraid he wouldn't tell us when to come in for gas. It was the most informal thing you could imagine."

In his pit, Chinetti watched Masten Gregory hurtle down the straight in the morning light. He looked up at the leader board and saw Gregory's #21 Ferrari, his entry, in second place. Emotions ping-ponged inside him. It was going to be a long day.

Over the years, Chinetti's relationship with Ferrari had grown complicated and at times contentious. Though it was Chinetti who convinced Ferrari to build his cars and sell them in America ("*You must believe that here sports cars will be a gold mine . . .*"), Ferrari had barely mentioned Chinetti in his memoirs, published three years before. There were fights over money. For Chinetti, business relations with Ferrari had been, in the words of Ferrari biographer Brock Yates, "an elaborate, protracted drama of artful parlays, suspended agreements, temper tantrums, operatic claims of impending bankruptcy, social ruin, family shame, incurable disease, and violent death."

Chinetti was having his best year ever in 1965. As a brand, Ferrari couldn't have been hotter. He would move about 240 cars before the year was out. But Enzo Ferrari was working new distribution lines in the United States, namely Nevada's gaming lord William Harrah out in Reno. Was Ferrari phasing Chinetti out? Many powerful players in the automobile industry believed that, if not for Chinetti, Ferrari's company never would have survived the postwar years. As Shelby—who knew them both—once said, "The Old Man cheated Luigi and generally treated him like shit."

How Chinetti longed to beat Ferrari on the track. Here at Le Mans, where it counted most.

At midnight, Chinetti's #21 entry was nowhere near the front. But by dawn, Masten Gregory and his teammate, the baby-faced Austrian Le Mans rookie Jochen Rindt, had moved up and were now stalking the leader. Their car was believed obsolete. The days when a year-old racing car could be competitive seemed long past; innovation moved too quickly. Neither driver had dreamed their car would even last this long. Gregory had raced so furiously in the opening laps, he pulled into the pit after his shift to find Rindt in his street clothes, shocked that the old Ferrari was still running. Even the mechanics were ready to quit. Why exhaust themselves and risk their necks when they knew the car would never survive? Gregory insisted and so Rindt changed into his driving clothes and took to the track.

All night long they'd raced on the ragged edge. They'd driven so hard they'd burned through their tires. Chinetti found himself, with flashlight in hand, hunting in the dead tire pile for Goodyears that still had some tread. By noon on Sunday, Chinetti's car was one lap behind the leader, a Belgian-entered Ferrari—293 laps to 292.

The morning's mildness turned to raw heat. The mood in Chinetti's pit was growing more tense with each lap, eyes darting back and forth between stopwatches, the Dutray clock, and the red Ferrari as it rocketed through the grandstands every four minutes or so. An announcement piped suddenly over the loudspeakers: *the front-running Ferrari had burst a tire on the Mulsanne Straight!* The crowd stirred. Chinetti retained his poise as the #21 Ferrari was moved up to first place on the leader board.

Some time later a Ferrari factory representative appeared in Chinetti's pit. He explained that the Belgian Ferrari, previously winning but now trailing in second place six laps behind, was on Dunlop tires. Chinetti's car was on Goodyears. Strategically, it would be best for Enzo Ferrari if the Dunlop car won, not the Goodyear car. Ferrari had a contract with Dunlop, the factory rep reminded Chinetti, and it wouldn't do if the Belgian Ferrari came in second due to a burst Dunlop tire. Surely Chinetti might enjoy a substantial discount on future cars for his showroom if Gregory and Rindt allowed the Belgian Ferrari to move back into first place.

"You're going to tell *them* to slow down?!" Chinetti responded incredulously. "How are you going to do that?" He was utterly offended.

Throw a race? Luigi Chinetti? At Le Mans!?

Masten Gregory drove the last shift terrified. He was miles ahead and now all he had to do was finish. His exhausted brain imagined clanking from the engine compartment and smoky odors. He had tortured this tired machine. Any second now it would seize up and die. It would have its revenge.

In the press booth, Jim McKay sat with Phil Hill doing commentary as a camera helicopter followed Gregory around the circuit on his last lap. A handful of cars escorted the leader in a traditional finishing parade. "These are literally tattered and battered oil-stained survivors," McKay said. "It's four o'clock right now. So as they come into our view this will be the end of the race. The crowd cannot see them as of now, except for those who are lining the roads along the countryside. This is one of the great moments in sports."

"Luigi Chinetti must indeed be happy today," Hill added.

When the #21 Ferrari crossed the finish line, the crowds surged forward onto the road. Gregory and Rindt climbed up on the podium, gendarmes surrounding them to keep the fans at bay. Standing in his black-rimmed Coke-bottle glasses, the Kansas City Flash looked like a more slightly built Clark Kent. He held up two fingers, making a V for victory. The rookie Rindt stood next to him, giving cameras the thumbs-up sign.* The racers were handed glasses of champagne and wreaths of flowers. (McKay: "It may seem odd to Americans to give these champions flowers, but that's how they do it here.") Nearby, Chinetti looked on, his normally sober face contorted with joy.

An American racing team had won Le Mans for the first time in history. It was a team of Ferraris, Luigi Chinetti's North American Racing Team.

The American national anthem played over the loudspeakers. In the crowd, Shelby stood sullen and tired. He saw his old friend Masten up on the podium. Shelby knew what it felt like to climb up on that podium a Le Mans champion, the thoughts that go through

---

* Jochen Rindt was killed at Monza five years later while practicing for the Italian Grand Prix. He became the first and only man ever crowned Formula One World Champion posthumously.

one's mind, like you could pick the globe up in your hand, toss it up in the air, and catch it. He saw Henry II's son Edsel II—who'd made the trip to France—approaching. Edsel, seventeen, had spent some time at Shelby's shop learning a few things. They'd gotten to know each other.

"Carroll," Henry II's son said, "what the hell happened? This is embarrassing."

When Shelby finally got a chance to congratulate the winner, Gregory summed up the story neatly.

"You know, that's a good car!" he said. "I stuck the clutch in that fucker and it revved up to 7,800 rpm! The thing didn't even bend the valves!"

Gregory was swept off to a victory party. Shelby headed to the nearest bar with the Goodyear tire guys to get himself drunk.

Ferrari's cars had won Le Mans for the sixth year in a row, taking the top three places in the last five. One Ford-powered car had finished, a Cobra in eighth place, six hundred kilometers behind the winner, a distance that equaled New York to Pittsburgh. Ken Miles called the race "the greatest defeat ever suffered by a team in the history of motor racing." "Murder Italian Style," read the *Sports Illustrated* headline. The fever that plagued the Ford 289 engines was diagnosed: defective head gaskets. And those 427s? The cars simply weren't ready to go. If there was any bright spot, it was Phil Hill's expert television commentary.

"He could wind up replacing Bert Parks on the Miss America Pageant," joked *Car and Driver.*

Henry II had spent, by one reliable estimate, $6 million (roughly $39.5 million today) to win one motor race in 1965. All he had done was help Enzo Ferrari sell more cars. It was said that Ford had spent all that money to strengthen Ferrari's reputation. People were comparing Ford's attempted invasion of Europe's racetracks to President Johnson's troops in Vietnam. The good intentions were there, but now the boys were stuck in a war they couldn't win, with no easy way out.

A month after Le Mans, Henry II received a letter from Rob Walker, heir to the Johnnie Walker whisky fortune who ran a powerful and well-respected racing team. Walker had fielded one of the Ford GT40s prepared by Shelby American at Le Mans. The car had lasted three hours. Walker was furious and humiliated, and he addressed his letter directly to Mr. Ford. "I was very disappointed," he wrote, "that Ford Motor Company should make themselves a laughing stock in European motor racing circles, which they undoubtedly did."

Henry II had heard enough. He called a meeting. He wanted Don Frey, Leo Beebe, and Carroll Shelby present. The three men found themselves sitting in Henry II's corner office. Mr. Ford's eyes bore into them. A journalist once described his gaze: "I felt like I was being unpeeled like an orange." Staring back at those blue eyes, Shelby knew there would never again be a man who would run an automobile company that big with that much power. The Deuce was the last of the corporate dictators.

"You got your asses whooped," Henry II said.

He passed each man a nametag. The tags said their names with one sentence printed next to them: "Ford wins Le Mans in 1966."

Shelby looked down, focused his eyes, and got the point. The year 1965 was now in the rearview. He'd get one more shot at Enzo Ferrari, and if he lost, Henry II was likely to pull the plug on the Le Mans campaign. History would record Shelby and Ford Motor Company as losers, and Ferrari would move on, unchallenged.

The three men left Henry II's office. Outside in the hallway, Frey stopped them. "I wonder what our fiscal responsibilities are?" he asked. They walked back in and asked Henry II, who paused, looking at them.

"You'd like jobs next year, wouldn't you?" he said.

"And that," as one member of the Ford racing team later recalled, "is when the shit hit the fan."

A couple of weeks after Le Mans, Shelby was asleep in his bed when the phone rang. A glance at the clock—nearly 3:00 A.M. It was Alan Mann on the line, an Englishman who was getting paid to oversee the Shelby American Cobra team at certain European races. Shelby had his hands full and he wasn't at the 12 Hours of Reims in France. A Cobra had placed fifth behind four Ferraris, but it had won the GT class.

The Cobras had clinched the GT-class Manufacturer's World Championship. Shelby might have gotten slaughtered at Le Mans, but he'd accomplished what he'd set out to do: beat Ferrari for the GT-class world title. It was the first time an American car company had won this vaunted award. And it happened on July 4, 1965.

Shelby hung up the phone and lay there in his bed. This had been his dream, and it had come true. But it was almost anticlimactic. So much had changed in the last two years. There were bigger dreams now, bigger battles. Many American fans didn't understand what GT-class racing was. All they understood was the checkered flag, and it was Shelby's job to bring it home.

The Magician of Maranello opened his newspapers after Le Mans and grew more furious with each article. Sure, a Ferrari had won, but it was Chinetti's car. His men and his machines had failed in the most high profile of races, and the papers were making sure everyone knew it. They loved him when he won, but Ferrari was so very arrogant. They let him know it when he didn't.

Ferrari knew Henry Ford II was going to come at him in 1966 with all the speed that his millions could muster. At his factory, Ferrari made his wishes known. They would build a new Le Mans car. Before the summer was out, his design engineers had completed the initial drawings. What was about to spring from those blueprints: the most striking car of its time, the Space Age racer defined. It would be the most powerful racing sports car to roll out of the Ferrari factory to date.

But who would be driving it? Fifty percent man, fifty percent machine. This was Enzo Ferrari's equation for victory. His lineup of talent was about to experience an unexpected shakeup.

There was no slowing down for Luigi Chinetti. Never one for interviews, he simply returned to America and went back to work. Glory and praise were not his emotional trophies. He was driven by a deep love of cars, and the thrill of the chase.

Chinetti was a master at scouting the next young talent. On September 19, three months after Le Mans, he debuted a new driver on his North American Racing Team at Bridgehampton, a talent who was about to explode on the scene like none other before him. He was twenty-five-year-old Mario Andretti, a tiny man originally from Montona, Italy, who had emigrated as a teenager to Pennsylvania, having been raised in poverty in a World War II refugee camp. At Bridgehampton, Andretti arrived for the race with a pair of red driving gloves, a gray helmet, and racing shoes with very thin soles, handmade in Sicily. He stood just five feet four, with an olive complexion and dark hair pushed straight back, Fonzi-style. He performed well in one of Chinetti's beat-up Ferraris. Chinetti promised him $500 for the outing.

"How many kids do you have?" he asked Andretti after the race, speaking in Italian.

"I have two boys," Andretti answered.

"Okay," Chinetti said, "I'll give you $500 for one boy, $500 for the other."

Andretti smiled. Good money. That Luigi Chinetti—what a guy.

The following week, John Surtees stood in the pit at Mosport Park, a racetrack northeast of Toronto, two miles from the Lake Ontario coast. The Canadian Grand Prix was two days away and the teams were practicing. Surtees was competing with the Lola team. He didn't make much money from his contract, and he had asked for Ferrari's permission to race with Lola. He could pick up on some of the highly innovative ideas the Lola engineers were working on and bring them back to Ferrari. The boss had replied, "In the classes we don't make cars for, okay." It was a tricky political move, and Surtees knew it. Some of Ferrari's inner circle didn't appreciate his working with another team. An *English* team.

The afternoon of September 24, 1965, was clear and crisp. Surtees watched his Lola teammate Jackie Stewart pull into the pit. Stewart stepped out of the car.

"It's not right," the young driver said in his Scottish accent.

Surtees stepped in, keen on figuring out the source of the Lola's ill temper. Sitting in the cockpit, he revved the engine and listened to the exhaust note. The V8 let out an angry snarl. He put the car in gear and motored onto the track. That's the last thing he remembered.

PART IV

THE RECKONING

SURVIVAL

AUGUST–DECEMBER 1965

> The Ferrari stands for all that is swift, virile, and enduring in automobile racing. Needless to say, the manufacturer who beats Ferrari can claim no little speed of his own, and the manufacturer who covets that distinction most ardently is Henry Ford II. This is the showdown year between Ford and his Italian antagonist. It began brilliantly . . .
>
> —*Sports Illustrated,* February 14, 1966

JOHN SURTEES CAME TO. He was lying in a hospital bed, tubes snaking into his body. He had no idea where he was or how long he'd been there. He could barely move and he was mummified, bandaged head to toe. He needed answers.

What had happened?

Where was he?

Was he going to live?

He swallowed hard. His wife Pat was there, looking at him with bloodshot eyes. She had been sitting there a long time. Surtees had been in and out of consciousness for four days.

The doctors had already briefed reporters on the driver's condition. He'd broken his spine and his pelvis, and he was in critical condition. "He was bleeding on the left side," officials at Scarborough General Hospital outside Toronto had said when the patient first arrived. "We hope there is no internal bleeding." But there was. When Surtees finally gained consciousness, his doctor explained to him that his pelvis had split in the middle, rupturing his kidneys. The pelvis had been ripped apart from the base of the spine. His left leg had been shoved four inches upward into his body, and his kidneys were still bleeding. He was not out of danger yet.

Surtees was in shock. He thought of a motorcycle accident his father had suffered that'd nearly taken his life, years before, during the war. History was repeating itself. Looking up at a doctor's cold face from his hospital bed, he felt utterly helpless.

"We don't want to open you up," the doctor said. "We want to wait and see if your kidneys seal themselves."

Which meant Surtees had to lie there and wait. There was time to think, too much time. He was tortured by the unknown. He knew he had crashed, but he had to understand: Was it his fault? Had he made a mistake? Mechanical failure? He had to know that he was incapable of a mistake, that he was a perfect driving machine. A Lola engineer who'd witnessed the accident arrived at the hospital and began to explain what happened on that practice day at Mosport. The engineer's face was pained and he spoke slowly. A racing engineer understood his job was to build the vehicle that a man was going to place 100 percent of his trust in, and in this case, the car had failed.

"I'm sorry, John," the engineer said. "You lost a wheel."

It *wasn't* Surtees's fault. He absorbed the horrific details. He was moving into Turn 1, a swift downhill right-hander, which placed almost all of the weight of the speeding car on the left front wheel. It had come loose. The car ploughed through a barrier and cartwheeled down an embankment. When it came to rest, Surtees had been pinned underneath and not responding. The track marshals that rushed to his aid had no idea if he was alive, but they knew gasoline was leaking and there was a chance of fire. They had to get him out fast. Surtees regained consciousness briefly in the ambulance, though he did not remember this. His wife and his mother and father-in-law arrived at the hospital the next day.

Friends and colleagues started showing up. When the Ferrari

team came to America to compete in the United States Grand Prix at Watkins Glen a week after the accident, team manager Dragoni and technical director Forghieri drove straight north to the hospital. Surtees was still woozy. It was no secret that he and Dragoni hated each other, but the team manager took a sympathetic tone.

"You just get my car ready for Mexico," Surtees rambled, though he wouldn't remember saying this. The Mexican Grand Prix was three weeks away.

"We'll wait and see," Dragoni answered, knowing there was no way Surtees was going to be leaving the hospital in the near future.

Enzo Ferrari phoned. Surtees heard affection in his boss's voice. Yes, he said, his side was severely banged up. Ferrari wanted to know, was he going to make it? Would he be able to return?

"I'll have a go," Surtees said.

Ferrari asked, "Which side is it?"

"My left. I'll be a bit shorter on one side when I get back."

"We'll build you an automatic," Ferrari joked, since Surtees was unlikely to be pressing down on any clutch pedals anytime soon.

"Thank you for your support," Surtees said and hung up. He soon learned that—even though he was not driving a Ferrari when he crashed, a fact that bristled the tempers of many in the Maranello organization—his hospital bills had all been paid. Ferrari's insurance was going to take care of them.

After more tests, the doctors told Surtees his kidneys had stopped bleeding. They were going to perform surgery to realign his pelvis. The operation was brief and successful. Dr. Paul McGoey, chief surgeon at Scarborough General, informed reporters afterward that the thirty-one-year-old Briton's injuries were stabilized. Dr. Frank O'Kelly added: "I would think he will regain full use of his faculties."

But would he ever race again?

The days went by slowly, visitors filing in, Surtees's wife Pat sitting through the hours with him. Pain set in. He could feel it heightening to peaks. Two weeks after the crash, his doctor told him they wanted him transferred for long-term treatment. "We are not really able to deal with your sort of injuries," his doctor said. Surtees had a choice: he could go to America or back to Europe. The American doctors would be likely to do more surgery, where in Europe, the thinking was to let the body heal on its own.

Surtees knew he needed no ordinary doctor. He needed a doctor who could understand the mind of a racer, a patient fully focused on returning to the sport that defined him as absolutely fast as possible. There could be no equivocation. He decided on Dr. Urquhart of St. Thomas's Hospital in London. Dr. Urquhart was the man who'd treated the great Stirling Moss after his 1962 crash at Goodwood, which had shattered nearly all of Moss's bones and left him near death. A phone call was placed.

"Don't let those Americans get their hands on you," Dr. Urquhart told Surtees. "They'll use the knife. Get back here as soon as it is safe to travel."

Surtees's friend Tony Vandervell, of Vanwall Formula One fame, arranged for the flight. The doctors told Surtees it was imperative that he remain still. Too much movement, and his kidneys could start to bleed again. Medics tied him tightly to a stretcher, and then carted him to the airport in a brand-new streamlined American ambulance on October 18. The big struts and automatic transmission made for a smooth ride. Vandervell paid for a whole row of first-class seats in a Boeing 707. The ambulance ride to St. Thomas's from Heathrow Airport wasn't as pleasant. In comparison to the American ambulance, this one was an English antique. The driver staggered through the gears as he maneuvered through the London traffic. It was agony, "an involuntary, and unwanted, vibro-massage," as Surtees described the experience.

At the hospital, Dr. Urquhart appeared. He looked down at Surtees and smiled. "Right," he said, "time's getting on. You're in a bit of a sorry state, but everything else is stable, so we've got to try and straighten you up. We've got to get your left side down a bit to match your right side. Do you see my assistant here?"

"Yes," Surtees said.

"He's big, isn't he? He plays rugby."

"Yes."

"I'm well built too, aren't I?"

"Yes."

"Well, we're going to take you down and put you on a table. He's going to get one end, and I'm going to take the other, and we're going to tug like bloody hell."

The next day Surtees was on the table getting yanked. The doctor on one side, the assistant on the other. They'd likely given Surtees enough muscle relaxants and pain relievers to floor a horse, but still—if this hadn't occurred in a hospital, it would've looked like medieval torture. At the beginning, Surtees's left side was four inches shorter than the other. At the end, it was a half-inch shorter, and that's how it would remain for the rest of his life. When it was done, the nurses came in. Surtees was so tired, he could barely open his eyes.

"We reckon you'll be working that leg on your own in about a week's time," a nurse said.

"Let's see if we can't do it in five days," replied Surtees.

He began exercising the top of his body by pulling himself up and down on a piece of rope slung over the back of his hospital bed, the stiff muscles in his arms and back swelling with blood. The nurses wheeled him in a chair down to the physiotherapy pool, which was actually a military water tank used for fire services during wartime, a simple construction of corrugated iron sheets bolted together. The buoyancy of the water allowed him to experiment putting weight on his left leg. After one hour in the pool, Surtees later said, "I felt as though I had driven the Le Mans 24-hour race single-handed."

On November 24, exactly two months after the accident, the big day came. A nurse arrived. "Today," she said, "we're going to try and get you walking on one leg."

Nurses handed him a set of crutches and helped him out of bed.

The pain was unbearable. Do not, under any circumstances, they told him, put weight on the bad leg. Surtees stood, holding himself up on the crutches. He took one step. Then another. He took a third step and then collapsed from exhaustion. The nurses caught him and pulled him back to the bed. "Feeling very poor and depressed," he wrote in his diary that day. The next day he managed eight steps. The next day he counted eighty-three. Dressed in a bathrobe over a pair of pajamas, his feet tucked into slippers and arms curled over crutches, he walked four hundred steps on November 28.

In December, the hospital's hallways filled with Christmas decorations, and Surtees's depression began to lift. He had begun the long, slow walk back to the cockpit.

At the Modena Autodrome, Enzo Ferrari invited the first members of the press to see his new racing sports car, the car that would take on America in the battle for Le Mans in 1966. It was still winter, the grass around the old racetrack a dull yellow, the gray Modena sky the color of lake ice. When Ferrari presented the car, the public saw the 330 P3 for the first time. On one side stood the man, on the other the car. They did not resemble each other at all. Ferrari was antiquated, wrinkled and ordinary in appearance. The car looked as if it could beat the Russians and Americans to the moon.

The so-called "P Cars" had developed one after the other during Surtees's time with Ferrari. The 250 P in 1963, then the following incarnations: the 330 P, the 330 P2, and now the P3. With each, the technology was more aggressive. Weight reduced while power increased. The new P3 had about 110 more horsepower than the original P car, and it was about 40 kilograms lighter. The shape was more aerodynamic. In profile the body was one continuous rising and dipping line, the nose arcing up into the front wheel wells, dipping down into the midsection, rolling into the wraparound windscreen, curving into the small of the back, then rising again into a wicked set of muscular hips. The entire package stood only 37.4

inches off the ground, even lower-slung than Ford's Le Mans car. It stood so low the driver would be situated almost horizontally, as if, cynics noted, one would be in a coffin.

"I have real confidence in the excellent work done by my staff," Ferrari told reporters. "This car will enter competition at Sebring."

The reporters circled the 330 P3 and jotted notes. Peering inside they saw a stark, functional cockpit. It was all business: black leather driver's seat, big tachometer front and center, leather-wrapped steering wheel with its Prancing Horse badge. Ferrari and his men went over the nuances. For the first time, they had equipped a Ferrari sports car with fuel injection. As for the engine, the car's heart and soul, they had stuck with the time-honored V12. At four liters, Ferrari's engine would be far smaller than the seven-liter Ford power plant.

"Four liters are enough because our experience tells us that this is the right capacity for a proper all-around balance," Ferrari said. "In fact, with an increase in capacity we would have an increase in [fuel] consumption and an increase in weight on the wheels, weight to move, and especially weight in braking. All these values create a lot of problems which are not compensated by the few kilometers per hour more that it is possible to obtain."

The V12 engine pumped out 420 horsepower on the test bed, less than the Ford 427 V8, which rated on the dynamometer in 1966 at 486 horsepower. But the Italian car weighed far less. On the track the Ford would be superior in top speed; it would outdo the Ferrari on the Mulsanne Straight. The Ferrari was more lithe and agile; it would spend less time in the pit refueling and would outrun the Ford in corners. The contest would be one of philosophy as much as muscle.

As Sebring and Le Mans approached, Ferrari was faced with the question of his number-one driver. Would Surtees make it back in time? If so, would he have the nerve to corner at 99 percent again? A bruised nerve was a fragile thing. Even if a man could physically step into a cockpit and close the door behind him, he couldn't know if he was fully healed until he hurtled into a tight corner at speed with cars an inch away on either side. Some at the factory were already working their own agendas, hoping they'd never see Surtees again. Ferrari knew there was no one in Maranello who could take *Il Grande John*'s place. He decided to gamble on his champion. He would sit back, watch, and wait.

18

REBIRTH

AUGUST 1965–FEBRUARY 1966

IN THE LATE SUMMER OF 1965, Henry II sent a card to the top executives in all the departments of his company. Each man who received this card would never forget looking down at it for the first time. The card had a Le Mans decal on it and a short message.

You'd better win.
—Henry Ford II

As of that moment, all those men knew their jobs were on the line. The failure of any car part at Le Mans would mean the failure of one man and the people who reported to him. If the transmission blew, then the head of transmissions would get hung out to dry. The same went for engines and foundry, brakes, and suspension.

The whole company was going to pool its resources and go racing.

Ever the champion of committee rule, Henry II's longtime confidant Leo Beebe formed a new group: the Le Mans Committee. It was a new regime, a task force made up of the heads of all of Ford's divisions. The group would meet every two weeks until the race. When the committee gathered for the first time in a conference room at the Dearborn Inn, a grand old brick hotel on Oakwood Boulevard five minutes from the Glass House, Beebe stood tall and slim, watching his colleagues file in—roughly twenty men, the engineering brain trust of Ford Motor Company. Don Frey, head of the Ford division. Roy Lunn and his team from Kar Kraft. Head of engines Bill Innes. Young faces and old. The men settled in and the room began to fill with tobacco smoke.

None of them had ever attended a meeting like this one. They were a group of executives—few of whom had the slightest experience in motor sport—charged with the task of building the perfect mechanical athlete and winning a historic race. Whatever they needed to do to beat that wily fox in Italy, they had to figure it out, and do it.

"Anything you want, let me know," executive vice president Charles Patterson told Beebe before the meeting. "We'll gold plate the gearboxes if necessary."

Beebe called the session to order. The onetime high school basketball coach had given his share of locker room speeches. Thus far the Ford Le Mans program, two years old, had been nothing but an embarrassment. In the automobile business, the launch of a car could be delayed if it wasn't ready. But they couldn't change the date of Le Mans—ten months away, June 18, 1966.

By the time the company men filed out of the conference room, a plan was set in motion. They were going to complicate the plot. There would be two Ford Le Mans teams,* one headed up by Carroll Shelby, the other by Holman Moody, who ran Ford's champion stock car team, the fastest NASCAR outfit going. John Holman and Ralph Moody knew how to make machines go fast. They were old-school stock car gurus, and their relationship with Ford Motor reached back to NASCAR's wild years in the 1950s. They were based in Charlotte, North Carolina, a hotbed of American speed talent. If Henry II sold cars on the heels of racing victories, Holman Moody had made Mr. Ford a lot of money over the years.

This arrangement would pit Shelby American against Ford's NASCAR boys, the glitz and sophistication of the Los Angeles sports car scene against southeastern stock car grit. Shelby's team worked with European-style sports cars on twisty tracks. Holman

---

* A third and less high-profile European Ford team was also assigned to the Le Mans project, headed by Briton Alan Mann.

Moody raced souped-up showroom Fords on oval tracks that had nothing but fast straights and left turns. Shelby obviously had the advantage, but if Holman Moody proved faster, Shelby would be humiliated. Which team would out-speed the other would be nearly as fascinating for the fans as would which team could beat Ferrari. Pitting their own men against each other—Ford Motor Company had stolen a page out of Enzo Ferrari's playbook.

Immediately after the first Le Mans Committee meeting, a group of Dearborn suits flew down to Daytona. It was seething hot that August morning when they walked onto the speedway infield. Ken Miles was already there. He eyed the suits suspiciously. They were going to do nothing but get in the way.

In the pit, Shelby's team manager Carroll Smith and chief engineer Phil Remington were making adjustments on a Ford Mk II. The suits stood idle, most of them having never been to a racetrack when the grandstands were empty. Inside all of their heads, these words were echoing: *You'd better win, Henry Ford II*. It wasn't just about competition or rivalry anymore, it was about survival. The moment Miles sparked the racing car's engine marked a new beginning, the beginning of the end.

Miles began to lap around Daytona's banked turns and infield road course, threading the car quickly through the hairpins, what he called "miserable, slippery little turns." The sun grew hotter by the minute, and Miles's body oozed sweat in the sweltering cockpit. Firemen and medical staff stood by. The company men stood there smoking their way through packs of cigarettes. They were learning a lesson about race car development: it required long and hopefully uneventful hours.

Miles would occasionally lose it in a turn and spin off the track, tires screeching. Then he'd motor back on, gaining speed. "I remember some really scary spins he had at Daytona," one Shelby man later recalled. "But they didn't seem to faze him."

After Daytona, the Ford car was flown to Dearborn for wind-tunnel testing. Then it was off to the company's proving grounds in Kingman, Arizona—ideal weather conditions during winter. Day and night Ken Miles lapped at speed, oversteering through turns, kissing 200 mph on straights. He could get a car so sideways he could practically see his own tailpipes. He moved from one hotel room to the next, living out of a suitcase, wearing the same old pair of socks, his old army jacket in desperate need of a wash.

Finally the car was getting the development work it had always needed, and with each lap, Miles's intimacy with the machine grew deeper. His colleagues described "an almost mystical sense" of a car's inner workings. The engine and gearbox were gaining durability, and the Shelby American crew was honing in on the most critical issue. The big 427 was so powerful and heavy, decelerating into corners was proving troublesome. Brakes functioned by converting the energy of motion into the energy of heat. A caliper grips a rotor like a squeezing hand, slowing its motion. The metal rotor spikes in temperature. (If a person were to squeeze a metal disc spinning rapidly, it would burn his fingers as it slowed.) When Miles's foot jammed on the brake pedal, the Ford's brake fluid instantly boiled and the half-inch-thick cast-iron rotors eventually shattered.

The crew moved to Sebring's flat airfield track to work on the braking system. With firemen and medical staff on hand, Miles moved around the circuit, trying out different setups and tires. Time and again team manager Carroll Smith logged brake failure on his clipboard. January 18: "brake fluid boiling," "complete loss of brake efficiency." January 19: "The car was uncontrollable at any speed under all conditions."

Throughout January, the team searched for solutions. They were making progress, but they ran out of time.

The Daytona Continental was the first race under Ford Motor's new regime, the Le Mans Committee. A number of high-up executives were watching closely. Shelby and Miles flew out of Los Angeles

International Airport on Saturday, January 30, 1966. It was a long flight, with stops in New Orleans, Tampa, and Orlando; there was plenty of time to think. The humiliation of Le Mans the year before still ate at Shelby. No amount of mouthwash could rid the bitter taste from his mouth. Another humiliation at Daytona would crush his reputation.

For the first time, the Daytona Continental was going to be a 24-hour race. Recognizing the exploding popularity of endurance racing in America, track owner Bill France stretched the 2,000-kilometer competition (about 13 hours) to a 24-hour marathon, hoping to cash in at the register, making it the only 24-hour race in the world besides Le Mans. Shelby feared the cars weren't ready for that long haul; brake failure at the wrong place and time, with the fans and reporters watching, could be catastrophic.

Ferrari's factory team wouldn't attend. Luigi Chinetti's team would represent the Italians. But Shelby's major rival was going to be the other Ford team, Holman Moody. The two teams would be racing nearly identical machinery. Shelby feared that the Dearborn suits were currying favor with Holman Moody. The Ford executive overseeing the Charlotte-based outfit, Jacque Passino, was a big NASCAR guy and a little shifty, and he held considerable sway within the company. Who knew what kind of politics or cash hand-outs were taking place behind closed doors? Holman Moody had signed major talent, most notably veterans Richie Ginther and Walt Hansgen, who were at that moment on their way to Daytona, too.

Another problem: Shelby had a long relationship with Goodyear tires. Holman Moody ran on Firestones. Goodyear and Firestone were locked in a bitter rivalry known as "The Tire Wars," and in Akron, Ohio—headquarters of both—the war was as intense as the Ford–Ferrari duel. Which rubber would prove superior on the track? Millions in sales hung in the balance. Ford and Firestone enjoyed a cozy relationship, back to the days when Harvey Fire-stone made tires for Henry Ford's Model Ts. In fact, Henry II's brother William Clay Ford had married Harvey Firestone's granddaughter Martha. The two companies were literally wed. Surely Dearborn executives had a preference of which Ford team would win at Daytona, Sebring, and Le Mans: *Holman Moody on Firestone tires.*

"Holman Moody this, Holman Moody that," Shelby complained. "You know," he told Carroll Smith, "someday you're going to get beat, and it better be by Ferrari."

Shelby checked into the Americana Beach Lodge, crashed for the night, then headed to the track in the morning. Already the world's top talents were filing into the speedway medical office for their physicals. "The driver list reads like a Hall of Fame roster," noted the *New York Times* in its race preview. In the pits, Mario Andretti was talking with Luigi Chinetti, the emerging star's eyes so dark and intense, they looked like black holes. Daytona was only his second sports car event.

"Mario," Chinetti said in Italian. "Everybody says you will break the car. Don't let me down, Mario."

"Just give me the lap time you want me to run, and I'll do it," Andretti said.

By the time the grandstands were filled and the pace car was leading the pack around the track, Leo Beebe had arrived in the Ford pit. He wasn't smiling. "We don't even know if our paint can go for 24 hours," he told a reporter. Ken Miles rolled along behind the pace car, headed for the start, while his teammate Lloyd Ruby lounged in the pit. Miles had won the pole; the aged mechanic and competition manager, who wasn't even supposed to be racing, had qualified faster than Andretti, Gurney, McLaren, and the rest. Miles had put all the work into this car, and now he was on the track, his fingers light on the wheel. He had no idea if and when he would lose his brakes. He had to sense a problem *before* it happened. He must have felt like a bullet in a gun barrel.

Twenty-four hours later, the crowds pressed down on the infield, where the winners were receiving their trophy. Shelby watched

Miles and Ruby field questions from reporters, their eyes glazed from exhaustion.

"We had confidence in our car," Miles said. "Some people told us the car wouldn't last, but it did."

Was he tired?

"A little," Miles answered. "I couldn't sleep very well last night. Some noisy buggers going around in automobiles kept me awake."

Miles and Ruby knocked nearly 10 mph off the average speed record at Daytona, an incredible feat given that the race had doubled in length. Their speed was astounding. One reporter called their victory "one of the most perfect drives in racing history." Another Shelby Ford finished second, a Holman Moody Ford third, and Andretti's Ferrari fourth.

For the first time in history, an American car had won a 24-hour FIA-sanctioned race. Even Luigi Chinetti—a naturalized American citizen—was pleased. When asked if he was bitter about his defeat, he answered in his Milanese accent, "I am proud for my country."

John Surtees pulled up to the Ferrari factory gate in a Fiat. The date was March 10, 1966. He stepped awkwardly out of the car, fumbling for his crutches. His wife Pat walked beside him as he hobbled onto the grounds. How many hours had she spent in pits at racetracks charting his lap times? And how many hours by his hospital bed? Now she was watching her husband battle against his body as he limped through the factory's large, metal, prison-like gate.

When they walked into the racing department, the technicians stopped their work. Surtees was a ghost, a man who'd come back from the dead. The English driver would always remember seeing tears drip down their faces at the sight of him. He looked over the machinery, his crutches sliding on the red tile floor. It was obvious that the Formula One project was well behind schedule. Enzo Ferrari's rivalry with Henry Ford II had taken center stage, and the focus in the racing department was the battle for Le Mans.

The technicians showed Surtees his "convalescence car," a Dino 246 Formula One car they had prepared for him, similar to the chariot that had carted him to his World Championship in 1964. Surtees's left leg was in such bad shape, he had difficulty climbing in. The mechanics pushed over one of the cranes used to move engines around the shop. Surtees let go of his crutches and grabbed hold. They winched him up off the ground and eased him into the cockpit. He wrapped his fingers around the wheel and placed his feet on the pedals.

Surtees and his wife sat down to dinner with Mr. Ferrari. There had been no attempt to replace the driver as number one on the team. Ferrari had faith that the champion would do everything in his power to return. They shared that singular focus. Who would either be without racing? Over dinner they talked about the new fleet of competition cars. Ferrari asked Surtees if he would consider taking on more responsibilities within the team. There was no talk of the Dragoni rift.

"Would you consider coming to live in Italy full time?" Ferrari asked.

Surtees liked the idea. He loved Italy, the people, the passion for speed, the way a win for Ferrari was a win for the entire nation. He could still recall the day he signed his first contract to race motorcycles for an Italian team—ten years earlier, in Count Agusta's office outside Milan. His racing career had for the most part been based in Italy since he was twenty-two years old. Ferrari offered him one of the flats he owned in Modena and Surtees accepted.

One morning in early March, the Englishman pulled into the parking lot behind the pits at the Modena Autodrome. Some technicians were standing around his convalescence car. When he climbed in, he must have struggled with his nerve. He knew the legendary story of the great Ferrari pilot Alberto Ascari. "You have to get straight back into the saddle after an accident, otherwise doubt sets in," Ascari said the day he climbed into a Ferrari at

Monza, still nursing wounds from a crash days before. Minutes later, Ascari was dead. Helmet and goggles on, Surtees stepped into the Dino. He fired the ignition.

Surtees's left leg was weak; it took effort to work the clutch. He moved slowly around those flat 1.5 miles at first, and then it all came back to him—the swift charge into a turn, the feel of the rubber's delicate grip on the blacktop. And that certain sound, the Ferrari engine, wailing inches behind his ears.

Each day, Surtees returned to the little circuit. It was early spring and the grass around the track was turning lush green. Mornings it was covered in dew, sparkling in the sun. When Surtees breathed in its fragrance, he smelled rebirth. As the days rolled by, reporters began to gather to watch him train, along with some locals, who sat in the bleachers that the Modena Automobile Club had erected. Surtees gained speed, accelerating deeper into his turns, lapping faster and faster.

On March 15, he shattered the Modena Autodrome's lap record, a coveted bragging right. All through the 1950s, Ferrari and Maserati drivers had battled to own this title, and it had belonged to some of the greats. Now it belonged to Surtees. The following day, he carved another .3 seconds off the record. On March 17, he went faster still.

News of the pilot's speed soon reached Ferrari's office. Ferrari expected nothing less of Surtees. He knew the gritty Englishman was battling against himself—his demons, his frailness. And he knew Surtees would win that race.

Still, Ferrari was preoccupied in March 1966. A number of dramas were unfolding in his office, each of them requiring his utmost attention. He was now entering serious negotiations with Gianni Agnelli, the grandson of Fiat's founder and its current chief executive. Ferrari was finally going to sell a large portion of his company. Long meetings were taking up a good portion of his time. Fiat was Italy's largest corporation. The Magician of Maranello knew that age was catching up to him. He was, after all, mortal. In the hands of Fiat and the Agnelli family, he could have faith that his work would continue after he was gone, and that the spirit of Ferrari would remain Italian. Already the two companies were laying the groundwork for their first joint project: a six-cylinder, two-seater customer car called simply Dino—the first customer car to carry this moniker, with no Ferrari, Fiat, or any other nameplate. The curvaceous little Dino sports car would eventually become one of the most coveted collectors' cars in the world.

That spring, Ferrari's band of trusted confidants began to dwindle rapidly. Vittorio Jano—the engineering maestro who'd created the Alfa Romeos that'd given rise to Ferrari in the prewar years, who sat with him by Dino's bedside in that fateful winter of 1956—was terminally ill and would soon take his own life. Enrico Nardi, revered Italian engineer who'd played a pivotal role in the Scuderia Ferrari's success in the 1930s, was growing sicker by the day and would soon be dead at fifty-nine. Ferrari's aged mother had died months before. That spring he learned that his long-time friend Battista Pininfarina was also dying. Pininfarina had designed the bodies of so many Ferrari cars that now graced roads from Japan to America. He would not live to see another summer.

Meanwhile, a young employee was beginning to make his presence known at the factory, and this, in some ways for Ferrari, was the most personal drama of all. The employee's name was Piero Lardi. Ferrari pulled young Lardi, who was twenty-one, into his inner circle. Lardi sat in on company meetings in the boss's office. During these meetings, Ferrari would sit at the head of a small and simple table facing the portrait of his son Dino, and ten to twelve men would gather around, elbow to elbow, most in suits but some in oil-stained jumpsuits. Lardi was put in charge of organizing the production of the new Dino 206 racing car (a cousin of the customer car mentioned above).

Some of Ferrari's closest confidants had known of Lardi for some time, but even to those who considered him a stranger, there was something familiar about his face: the aquiline nose, the structure of the cheek bones, and the strange brooding eyes. Piero Lardi looked curiously like his boss Enzo Ferrari. The old man had kept a secret for so long, and now it was creeping out. The mistress, the bloodline.

Ferrari had another son.

Piero Lardi was born during World War II. At the time, most young men were in uniform and in many cases, women filled their places in industry. These were the years that Ferrari built his factory, and many of his employees were initially women. One such young local named Lina Lardi had drawn the affection of her boss. They had an affair. Through all the years since, they had retained a clandestine relationship; Lina now lived in an apartment in Modena owned by Ferrari, along with her son, and Ferrari made frequent trips to sit and visit. By 1965, Piero was getting the first taste of his birthright. That his first major project was a Dino, a car named after Ferrari's deceased legitimate heir, may or may not have been a coincidence.

Amidst all these weaving plots, the Ford Motor Company juggernaut loomed. The new Ferrari competition cars would debut at Sebring on March 26; Surtees would not make the trip. Ferrari sensed that the Americans were mighty, that their resources were many times greater than his own. Through the years he had funneled all the money from his customer cars into his racing. But the Italian *lire* against the American dollar in 1966, against Henry Ford II's bottomless pockets—could a man like Ferrari win in such a duel, no matter his genius?

Ferrari's negotiations with Fiat were becoming increasingly critical. The days when a small, independent company could build racing sports cars and compete against major corporate powerhouses were coming to a close.

The week before Easter, Ferrari approached his deputy Gozzi. "Sunday there's Sebring," he said. "And Wednesday practice for Le Mans."

Gozzi saw where this was going. "I don't feel very well," he said, "but if you need me I'll be at home and available." Gozzi had spent Easter the year before at a race in Sicily. He wanted to stay with his family.

"Sort out your health," Ferrari said. "It is not necessary that you go to Le Mans, but you must see Sebring. We'll lose the race. But I'm interested in knowing firsthand how the Dino and the new P3 go."

19

BLOOD ON THE TRACK

MARCH–APRIL 1966

HENRY FORD II stepped through his private executive entrance in the Glass House one Friday morning, his face grim. Waiting for him were two of his vice presidents, whose faces were even grimmer. They had a major problem on their hands.

A tornado of controversy had touched down in the nation's capital. A young activist named Ralph Nader had published a book called *Unsafe at Any Speed* on December 1, 1965. Its message: Automobiles were killing off Americans at the rate of nearly 48,000 a year and that number was rising fast, and car manufacturers were to blame. Fueled by greed, Nader claimed, they peddled the drug of speed and style, ignoring safety altogether. *Unsafe at Any Speed*'s tone was extreme, so much so that it read like a document of religious fanaticism from the first page: "For over a half century the automobile has brought death, injury, and the most inestimable sorrow and deprivation to millions of people. With Medea-like intensity, this mass trauma began rising sharply four years ago reflecting new and unexpected ravages by the motor vehicle."

Who was Ralph Nader? No one in Detroit had heard of him. A lawyer by trade (Harvard Law School, class of 1958), he was a lean six-foot-four thirty-two-year-old of Lebanese decent, a bachelor who lived in an $80-a-month rented room in Washington, D.C. He didn't own a car. A couple years earlier he had abandoned a small Connecticut law practice and had hitchhiked to the nation's capital. He had gotten a job as a consultant to the U.S. Department of Labor and started writing freelance articles about public health issues, one of which eventually turned into *Unsafe at Any Speed*.

Prior to Nader, the debate over highway safety focused on whether lap belts should be standard in new cars and on a new technology called "the air bag." In February 1966, two months after the book was published, President Johnson called the crisis of death on highways America's gravest problem outside of Vietnam. In an era when millions were suddenly confronted with a mistrust of government, big business, authority of any kind, Nader posed the question: Should Americans mistrust their own cars?

Henry II hadn't read the book. He didn't have to read it to know what it said, and to know that it named names. His, for example. And Lee Iacocca's, Don Frey's, Roy Lunn's. The book focused primarily on the Chevrolet Corvair, made by General Motors. But for Henry II, the automotive executive who'd invested the most in speed and racing as a marketing tool, who had himself shredded the Detroit Safety Resolution four years earlier in the spring of 1962 (exactly the year that, Nader claimed, highway deaths began to skyrocket), the safety controversy proved particularly sticky. Nader's book caused such outrage, Senator Abraham Ribicoff of Connecticut had organized government hearings in the Capitol. In a few days' time, Henry II's vice presidents were going to make statements before a panel that would include U.S. Attorney General Robert Kennedy, a group of senators and congressmen, and countless television cameras and reporters.

In the Glass House, Henry II huddled with his vice presidents, trying to script a statement. What was their policy on improving motorists' safety? Was speed and style in fact the enemy? Henry II listened to his advisors' statement and he grew furious. It was all wrong.

"If you take this to the Senate committee they're going to laugh you out of the hearing room," he said.

Yes, he had a major problem on his hands. And it was about to blow up in his face.

Three years into Ford Motor Company's campaign to defeat

Enzo Ferrari, no driver had perished or been critically injured at the wheel of a Ford Le Mans car. As the 1966 season rolled on, and the Nader situation gained more publicity, the company's luck was about to run out. The timing could not have been worse.

Carroll Shelby stood on the pit wall at Sebring angrily waving a hammer at Ken Miles, who was cranking past at high speed. It was late in the 12 Hours of Sebring, which had started at 10:00 A.M. Darkness had fallen. Dan Gurney (in the #2 Ford) was in the lead, but Miles (in the #1 Ford) was hunting him down. They had team orders: *no racing each other.* It didn't matter which Shelby American Ford won, as long as one did. They were ignoring pit signals to slow down. Miles saw Shelby waving that hammer and he backed off. When he passed the pit after his next lap, he gave Shelby the finger.

The race had been a brutal, violent affair. Canadian racer Bob McLean had been killed in a Ford GT40. He'd lost it in a corner and had bludgeoned a telephone pole, his car exploding in a fireball in midflight. A collision between Andretti's Ferrari and a Porsche had sent the latter car hurtling off the track, killing four spectators. (Andretti made it back to his pit after this shunt, where his car erupted in flames. He left the track immediately and did not learn of the spectator deaths until the following day.)

Gurney pulled in for a final pit stop before the finish. He huddled with Shelby while mechanics refueled the tank and checked tires. Behind Shelby, the electronic leader board showed the positions. It was Gurney, then Miles.

"Well," Shelby told Gurney, "you've got it won. Go out and take it easy."

At 9:59 P.M., the crowds pushed toward the pit row to see the checkered flag wave. The reporters had already tapped out their leads: *Gurney wins. Five killed at Sebring.*

Suddenly, Ken Miles rocketed over the line in first place, taking the flag. Gurney followed on foot, pushing his car and panting. His engine had given out a couple hundred yards from the finish.

Shocked, the crowds screamed for the winner and applauded for Gurney, who was dumbfounded and heartbroken. Someone ran into the showers and found Miles's teammate Lloyd Ruby soaping himself.

"Lloyd! You won!"

Miles and Ruby were victorious, once again in record speed. But the outcome had left Leo Beebe a little miffed. Miles had ignored team orders. Had Gurney's engine blown because Miles had forced him to keep his foot down?

A bleary-eyed Franco Gozzi hung his head and left for his hotel. He faced the task of informing Ferrari of the news. The new Ferrari 330 P3 had not finished the race; its transmission had given out. And the new Dino racer had fared no better. When he reported back to Maranello, his boss sounded more than a little irritated.

"Yes, yes, yes," Enzo Ferrari said. "I see. What a nice Easter you gave me."

A handful of Ford cars were airlifted to France, and just days after the 12 Hours of Sebring, Ken Miles was on the track at the Le Mans test weekend. Ford Motor Company's aerospace division sent a crew to install state-of-the-art computer equipment in his Mk II. As the car burned fuel, the jumble of wires, plugs, and panels sucked in data.

On the Mulsanne Straight, Miles cracked 200 mph. He hit 202, 203, 204. It had rained in the morning and there were water hazards on the pavement. When Miles stomped on the brake at the Mulsanne hairpin, engine speed spiking as he downshifted, an oscillograph spit out a stream of graph paper tracking his rpms. The readout looked like an electrocardiogram of a man in the grip of a coronary.

Miles could feel it: *this was his car, his time*. At the unlikely age of forty-seven, his glory moment was about to arrive.

A few yards away, Surtees was preparing himself for a run in the new 330 P3. The red car, with its aggressive, almost belligerent profile, drew plenty of attention, but all talk among the journalists wandering the pits was the speed of the Ford cars. Surtees was still limping; he had yet to enter a race following his hospital stay. This murderous track would be as much a test of the driver as it would be the car. His nemesis Eugenio Dragoni approached him, concerned by all the talk of the fast Americans.

"John, you've got to do something for me," Dragoni urged. "You've got to go out there and set the fastest time. You're the only one who can do it."

Surtees slipped on his helmet. As he eased onto this familiar billiard-table-smooth pavement, he hammered down and the engine shrieked. He headed for that long, storied 3.5-mile straightaway.

Carroll Smith stood in the pit, logging lap speeds for the official Shelby American report. He saw a Ford Mk II jet past, driven by the veteran Walt Hansgen. Smith looked at his stopwatch and shook his head. It read 3:46.8. All morning long Hansgen had been told to slow down, but the driver kept lapping faster and faster. He seemed intent on proving he could match Ken Miles's speed. Smith watched him barrel down the pit straight and saw the car's tail suddenly wiggle. Hansgen had hit some water and was hydroplaning. Tires screeched as the car fishtailed, the driver obviously working frantically, making split-second corrections at well over 100 mph.

"*Christ,*" Smith groaned through clenched teeth, *"I hope he can sort this one out."*

The Ford hit a sand bank and cartwheeled end over end, landing in the dirt in a cloud of smoke. The pits emptied. Spectators and reporters rushed to the scene. The wrecked car no longer resembled an automobile except for the wheels sticking out on bent axles. Inside, Hansgen was trapped, and possibly alive. He was unresponsive. The crowd continued to grow; hundreds had gathered. Fingers pointed and camera shutters clicked. Was this what they had come to see?

At a nearby hospital that afternoon, a British Ford executive, Walter Hayes, stood near the operating room for what he would later call "the longest hours of my life." Hayes was Henry II's closest confidant in Europe, an executive who called Mr. Ford "Henry," and was the only man Henry II had invited to his wedding the year before. Hayes was a towering figure in the European automobile industry. He volunteered to accompany the injured driver from the track to the hospital, as he spoke French and could communicate with the doctors. A surgeon walked out of the operating room.

"Did you know him well?" he asked.

"No," Hayes said. He explained that he had met Hansgen that morning.

The doctor shook his head. "Those who come to Le Mans know what they risk," he said.

At the Le Mans trials in 1966, John Surtees set the fastest lap time. Walt Hansgen died five days later in a U.S. Army hospital in nearby Orleans.

On April 13, 1966, Leo Beebe called a Le Mans Committee meeting to order at 12:30 P.M. in Conference Room B at the Dearborn Inn. The group arrived promptly and took their seats. On the agenda: the Le Mans test weekend. A Ford executive who'd attended the trials took the floor and recapped in detail the crash that had killed Walt Hansgen—weather conditions, the wet-weather tires Hansgen was on, the fact that practice continued soon after the accident. Twice before he crashed, Hansgen had been signaled into the pit and ordered to slow down. Carroll Smith told him to lap no faster than 3:50, but Hansgen wouldn't listen. "Walt's subsequent laps were 3:59, 3:48.5, and 3:46.8," the executive reported. Lap twenty-one had ended Hansgen's life.

The conversation turned to the next item on the agenda: the final preparation for Le Mans. The race was two months away.

THE BLOWOUT NEARS

MAY–JUNE 1966

> "You are the most completely egotistical bastard I've ever met."
> "You don't understand. When I go in there, if I don't really and truly believe I am the best in the world, I had better not go in at all."
>
> —Conversation between writer Ken Purdy and a famous bullfighter

ENZO FERRARI STOOD in the paddock at Monza with one of his aides. It was practice day for Monza's 1,000-kilometer sports car race. The event would be a significant test for Ferrari's new Le Mans car. It would also mark the return of John Surtees to competition. Dark clouds rumbled and the place buzzed with intrigue. In practice, no one could catch Surtees.

Ferrari spotted John Wyer on the grounds. He knew Wyer; everyone knew "Mr. Aston Martin." Wyer had led the Ford Le Mans effort early on but had since faded into the background. He was at Monza with some backup Fords that weren't expected to pose any threat, and he was standing with another man who was clearly from Ford Motor Company. Wyer approached Ferrari, whom he called "The Great Man," and they shook hands.

"I'd like you to meet Mr. Ray Geddes," Wyer said, nodding to the Ford executive next to him.

Ferrari turned to the Ford man. It was the first time since the failed business deal of 1963 that he had met a Ford executive from America face to face. The American launched into a formal greeting.

"I would like you to know, Mr. Ferrari, that we at Ford have a great respect for you."

Ferrari's aide translated and the great man responded in Italian. "Yes, I know," Ferrari said. "Like America respects Russia."

He walked away.

Surtees and the Ferrari engineers had clashed over the car in the days leading up to the race, a vicious row that had left tempers seething. The 330 P3 had been built and developed while Surtees was laid up in the hospital and he wasn't happy with it. The car oversteered, he complained. The aerodynamics were off. The engineers who'd done the development work were furious. Surtees had an insatiable desire to think technically, improve, make faster, and he seemed incapable of compromise. "John's idea of the perfect team is one in which Surtees is the owner, Surtees is the designer, Surtees is the engineer, Surtees is the team manager, and Surtees is the driver," one critic said of him. "That way he could be certain of one hundred percent team effort from his staff!"

The pilot had to believe his instincts were absolutely correct. He couldn't afford to doubt himself. Ferrari's engineers did what they were told and by the drop of the flag, Surtees and the car were one. He drove a masterful race in foul weather at Monza, losing his windshield wipers along the way. In the rain, in his first race back, Surtees took the checkered flag. He and his teammate Mike Parkes pushed into the lead on the first lap and never let it go.

As June approached, Ferrari and his team conducted a final shakedown at Monza. As drivers looped around and around in the 330 P3, slicing into the *Curva Grande* and the *Curva Lesmos,* down the blistering back-straight toward the *Parabolica,* Ferrari strolled the grounds. Aside from his home track in Modena, where few actual races were held, this was the only racetrack he'd seen with his own eyes in years. The sight of the empty grandstands summoned memories of his early days when he was a young competitor. Nuvolari's spirit could be felt here. God only knew what kind of heroism the Flying Mantuan would have demonstrated to make sure no American car defeated the Italians.

Not since the days of Nuvolari and Hitler's "silver arrows" had Ferrari experienced the kind of international rivalry that faced him in 1966. Both he and Henry II lived and breathed their companies. As Vittorio Jano once said, "[Ferrari] has no other satisfaction. His family, his very life, is that creature of his, La Ferrari." As for Henry II, there was no separating Mr. Ford from Ford. "He never, in sickness and health, in public or in private, discerned any separation," wrote Henry II's friend and biographer Walter Hayes. "The company always came first, ahead even of family." Now this rivalry between two industrialists had become one between two nations, two continents.

Everyone wanted a piece of the action—tire companies, spark plug manufacturers, gasoline companies. Rumors hit the papers that the French Ministry of Interior was considering canceling the race due to safety concerns. Meanwhile, big-budget movies featuring both the Ford and Ferrari teams were about to hit the silver screen. Ferrari was the focus of John Frankenheimer's fictionalized *Grand Prix,* while the Ford Le Mans car and the Mustang enjoyed a starring role in the Golden Globe–winning French film *Un Homme et Une Femme.*

The two industrialists made gestures of respect. Ferrari had received a note from Henry Ford II. It was a greeting card paying respect, a sporting gesture from America before the ultimate showdown. Ferrari returned with his own note, a few simple lines.

Ferrari had an acute talent for using public opinion to his advantage. He was a master of manipulation and subterfuge, and in this new era of mass media, he would put his skills to work. He published an article in the Italian magazine *Autosprint* accepting defeat before the race began, knowing the bit of gossip would get picked up by the international dailies.

"We know that nothing is being done to resist the steamroller of the Americans, who will find the road open to success in sports car racing," Ferrari wrote. "We fought on the track with autos and at the table against the abuses of power in the regulations. Even while continually winning races I understood that we were gradually losing them. We intensified our activity to the utmost, but we managed simply to slow down the approach of the steamroller. The battle was lost in advance."

Ferrari willingly cast himself as an underdog. Ford was Goliath, and he David. If he lost, well, he had predicted defeat. Henry II had done nothing but *buy* Le Mans with his countless millions. And if Ferrari won, as he absolutely intended to? In a lifetime defined by tragedy and victory, perhaps he would achieve his greatest moment.

"Incompetent dictator!" Surtees shouted.

"Ill mannered!" Dragoni fired back. "Untrustworthy!"

The pair faced each other in the pit at Monaco. The date: May 21, a few days after the Monza sports car race and less than a month before Le Mans. There in the shade of some pine trees, the day before the Monaco Grand Prix, Surtees and Dragoni finally clashed, firing threats at one another in plain view of reporters and mingling racing fans. Dragoni sat on the pit wall staring at Surtees through his thick-framed glasses, his back hunched. Surtees was in his racing coveralls, hands on his hips, teeth clenched. Gozzi stood by, stunned and helpless. He'd never seen anything like this occur before. Bitter words "hit me like whiplashes," Gozzi later recalled. "I couldn't believe it. It seemed impossible to me that such a scene could take place in the team—especially during practice for a Grand Prix when maximum concentration is needed."

Dragoni had planned to place Surtees in a new 12-cylinder F1 car, but the Briton didn't want to drive it. He argued that number-two driver Lorenzo Bandini's car, a 6-cylinder, was faster. Surtees was number one so he should get the best car. He called the new 12-cylinder "gutless."

"We make 12-cylinder road cars," Dragoni spit, "so you've got to race the V12."

"I told you, I cannot win with it. I thought we came to Monaco to win the race!"

Dragoni shrugged. "Oh, you'll win the race, all right."

Gozzi understood this was more than an argument. It was two seasons' worth of ire boiling over. And there seemed no way to cool off either man. The security of the team was at risk.

The following morning, Surtees relented and drove the 12-cylinder, firing off the line in the new car. He hurtled through the ancient city streets, past Monte Carlo's famous casino and the Hotel de Paris, down to the harbor and the tunneled straight that coursed along the sea. He led for thirteen laps. Then his differential failed. Helmet in hand, he headed back to his hotel without talking to reporters. Gozzi believed that, in Surtees's paranoid state, the driver suspected the mechanics of sabotage, that they'd caused the problem on purpose to knock him out of the race.

On Monday, Ferrari summoned Surtees, Dragoni, and Gozzi to his office. By this time, the boss had gotten wind of all the details—the argument, the mechanical failure. Right there in the boss's chamber, Surtees and Dragoni went at it again. Surtees was incensed, and Dragoni—seeing an opportunity to get rid of the Englishman once and for all—pounced.

Ferrari had heard enough. "We will decide what to do after Le Mans," he said. He called an end to the meeting and it seemed neither party had won; it was a draw. But, unquestionably, something had to be done, or things would continue to unravel. Later that evening, Ferrari called on Gozzi. He'd done some thinking. Surtees's place on the team was growing more precarious by the day, and Ferrari needed a backup plan. The Englishman was not like the others; he couldn't be controlled. Ferrari had his eye on the new talent from the United States, the fast kid being groomed by Luigi Chinetti.

"Contact Mario Andretti in America," Ferrari told Gozzi. "Immediately."

Meanwhile, in Dearborn, the Ralph Nader problem had spun out of control.

All spring, the *Detroit News* slapped onto front stoops in the posh suburbs each morning, one headline more shocking than the next. Paranoia had taken the world into its grip and the fist was squeezing. UFO sightings, race riots, druggies, and the nuclear threat. The Vietcong, pinkos, freaks, and spies. The Russkies had launched the first satellite to orbit the moon. "Subversive" rocker John Lennon had claimed his band "was more popular than Jesus." Over breakfast tables in many homes around Detroit, however, it was the scandalous Nader stories that opened eyes the widest.

Senate hearings over the safety issue had exposed what many believed an ugly truth about Motor City, best expressed by a pithy maxim attributed to Lee Iacocca: *Safety doesn't sell.* (In truth, it's unlikely that Iacocca said exactly that.) A backlash from Detroit caused Nader to fear for his own safety. He charged that he was being followed, harassed with midnight phone calls, his friends sought out and grilled by mysterious operatives. ("Are you asking me if he is a homosexual?" one Nader confidant said when mysteriously questioned by a private eye. "Well," came the answer, "we have to inquire about these things. I've seen him on TV and he certainly doesn't look like . . . But we have to be sure.") Nader even claimed that attractive young female strangers had tried to lure him into seedy situations. Were the Detroit companies behind these strange games?

The FBI got involved. Nader used the publicity to further his cause. So ambitious was he, it was as if this previously unknown lawyer had his eyes on the Oval Office.

The debate went global. Writing in the French magazine *L'Europeo,* Enzo Ferrari defended his passion against the safety crusaders: "A thorough survey on this thorny subject shows that most accidents occur for reasons which have nothing to do with speed," but rather careless driving, he claimed. As the Senate hearings moved forward, Congress drafted what would soon become the National Traffic and Motor Vehicle Safety Act of 1966, the first instance in history of government regulation of the car industry.

Henry II was furious. Nader's crusade was misguided, misinformed. Meanwhile, new car sales were plummeting. The Deuce had to take a stand.

"You will agree that we are being attacked on all sides, and we feel these attacks are unwarranted," Henry II said in a speech that was excerpted in international papers. "Naturally, when 50,000 people a year are killed on the roads of the U.S., this is a bad situation. On the other hand, to blame it solely on the automobile is very unfair." He paused. "We have a fellow called Nader. Frankly, I don't think he knows very much about automobiles. He can read statistics and he can look up a lot of facts that are in the public domain and he can write books, but I don't think he knows anything about engineering safe automobiles. I hope Congress will consider the problems that they may force on the automobile industry in depth before they pass a law. If they do something irrational, they can upset the economy of this country very rapidly."

Once again, "safety" had become the buzzword in Detroit. It had come full circle. Only now automotive safety wasn't just a buzzword, but a moral and political contest of wills that captured the attention of Capitol Hill and the world. As the scandal grew more ominous with each day, Ford Motor Company was entering the final preparations in a highly publicized quest to win what many believed the most dangerous sporting event of any kind. The Le Mans campaign was speeding right into the face of the safety crusaders. But the company had come too far and spent too much to turn back. In fact, Henry II had been officially invited to the thirty-fourth Le Mans *Grand Prix d'Endurance*. He was going to serve as Honorary Grand Marshal.

In laboratories throughout Dearborn, employees of all kinds were working to meet deadlines, to be assured that their specific job was perfectly executed. Every eventuality at Le Mans had to be considered and addressed. There was talk of nervous breakdowns and requests for early retirement. All the races, test sessions, debriefings, and Le Mans Committee meetings had led to this: the final build of a small fleet of racing cars. This work, and the theoretical lab work that accompanied it, represented probably the most sophisticated study of the inner workings of an automobile that had ever been undertaken.

Engineers in one lab were studying air intake, how the car breathed. By making alterations to the combustion chambers in the 427 engine so minute they were invisible to the human eye, the team unearthed 35 more horsepower. In another lab, electrical engineers were designing the final wiring system. The schematic utilized no moving parts, nothing that could be damaged by extreme vibration. The wire's insulation protected electricity flow in temperatures as high as 275 degrees Fahrenheit. All switches, light bulbs, and wires came from heavy-duty Ford truck parts bins. The windshield wiper systems were the same as those used on Boeing 707 aircraft, geared to sweep at speeds of 105 to 115 times per minute.

Transmission engineers calculated every gearshift. A team of two drivers would spend a total of 3.6 hours in the act of changing gears—some nine thousand shifts. Inside Ford's Reliability Laboratory, brake experts were trying to solve the most critical problem. Using mathematical equations—complex calculi scrawled across page after page—the team found that the Frisbee-sized brake rotors had to absorb a total of 12,597,900 ft/lbs of kinetic energy every lap. When the driver hit the brakes at 210 mph at the end of the Mulsanne Straight, the cast iron would spike in mere seconds from ambient temperature to more than 1,500 degrees Fahrenheit. The team experimented with vents and substances that could strengthen the cast iron, such as chromium and molybdenum. Both Shelby American and Holman Moody were working on a new brake rotor system that a skilled pit crew could change in a single minute.

In test room 17D of Ford's Engineering and Research complex, the sound of an engine's screams echoed against cement walls and down a hallway. Technicians had named room 17D the "Indoor Laboratory Le Mans." Inside, a 427 Ford engine was laid on a special new test bed. Made by General Electric, the test bed cost many millions and it was rigged with a sophisticated computer.

Using the measurements from the oscillograph mounted in Ken Miles's car at the Le Mans trials in April, the team programmed into the computer the engine speeds and gearshift patterns of a single lap on the circuit. The computer could then "drive" through a near-exact simulation. All attempts were made to make the computer and other equipment behave like a human driver. Gearshifts were actuated by pneumatic switches, each shift programmed to take between .3 and .7 seconds, mimicking Miles's arm-throw during competition. The switch that moved the clutch was set to press it .25 to .5 inch past the declutching point, just as Miles's left foot would in the cockpit. So elaborate was the test bed, it could account for the effect of cornering and wind resistance on the performance of the engine, even as it stood still.

Sitting at a console that looked straight out of *Star Trek* (which debuted on NBC that year), a technician threw a switch that kicked the experiment into gear, and the engine began to race through phantom laps. Every two hours, the technician hit a cycle switch that simulated a pit stop. It shut the engine down momentarily as a phantom pit crew jumped into action. And then it was back onto the track, shifting up through the gears and hammering on the throttle.

Day turned into night and night into day, but the cosmos didn't register in the Indoor Laboratory Le Mans. The same fluorescent lights beamed and the clock on the wall ticked through the hours. Every time the engine failed, a full inspection followed, adjustments were made, and the race began again. The experiment didn't end until the engine could survive the abuse not of one 24-hour race, but two in a row.

All this work took place behind closed doors, outside the glare of the media spotlight. The media focused on an emerging star, the man favored to drive this American car to victory.

That spring, Ken Miles attended a cocktail party at the British Consul General's home. As he smiled and answered questions, wrinkles cut deep into his angular face. He was enjoying celebrity for the first time in his life. He had never courted fame, only success. But the two came hand in hand. One guest asked naively if he could ever become a race car driver. Miles looked deep into the stranger's eyes, as if the answer lay there.

"That's up to you, sir," he said. "Isn't it?"

"There were some British movie stars at the party," recalled one man present. "But it was Miles who captivated me. The movie stars had their little groups of admirers, but Miles had me and several others, who wondered about this man whose life was dedicated to speed."

A growing fan base got a glimpse into the very private man's personal life. Everything Miles did, he seemed to have an incredible passion for: gardening, fine wines. He had a cat that he had taught how to defecate on the toilet. All spring he'd been racing against the best in the world—Formula One racers—and beating them. "Miles the Man for Le Mans," read a *Los Angeles Times* headline. "Ken Miles and [driving teammate] Lloyd Ruby have pushed American prestige to the peak," read another.

"I am a mechanic," Miles described himself. "That has been the direction of my entire vocational life. Driving is a hobby, a relaxation for me, like golfing is to others. I should like to drive a Formula 1 machine—not for the grand prize, but just to see what it is like. I should think it would be jolly good fun!"

His ultimate goal was to stand on that podium and sip the victory champagne at the greatest of all races. "I feel our chances at Le Mans are very good indeed," he said days before departing for France. "These cars were built for Le Mans." With his teammate Lloyd Ruby, Miles had won Daytona and Sebring in 1966. Le Mans would make for an endurance racing triple crown, something no man had ever achieved.

Mario Andretti was in Indianapolis when the phone call came. It was a call that the twenty-six-year-old would never forget. Franco Gozzi was on the line from the Ferrari factory. Had Andretti any plans? Could he, maybe, head to the airport? Right now?

That very week, Andretti had gone from hot young driver to an international star, the talk of the sports world. In his second year at Indianapolis, he outran everyone in qualifying trials, winning the Indy 500 pole with a record speed of 165.899 mph. His soaring performance had stunned all the Gasoline Alley veterans. They'd driven furiously, trying to keep up with him. Driver Chuck Rodee had crashed and perished in the process.

Andretti was what the sport of racing was begging for in the new age of television and wildly increasing popularity: a young idol with a handsome face, stars in his eyes, and a full tank of attitude. He had grown up in Italy and, once, in 1954, he had gone to the Grand Prix at Monza. The trip was quite an extravagance; the family had very little money. Andretti saw Ferrari drivers defend Italy's honor against the Mercedes-Benzes. He was fourteen years old. His family emigrated to America one year later. They spoke no English. Mario and his older brother Aldo started racing a Hudson Hornet on dirt tracks in 1959. Short of funds, they shared the same gray helmet. Aldo crashed and spent two weeks in a coma, never to compete again. Andretti continued to wear a gray helmet in his brother's honor.

In 1964, Andretti became an American citizen. He called that moment the greatest day in his life. But he'd never forget his roots: his lust for speed was born at Monza. He had a dream of returning to Italy and competing in Ferrari colors. But was he ready to step into that cauldron?

"Please try to remember me a year from now," Andretti said to Gozzi in Italian. "I don't feel that I can do the proper job. I'm not experienced enough yet. I'll do everything possible to gain experience." He was all but begging Gozzi to leave the door open, and indeed, Andretti's time in Italy would come.

At the factory in Maranello, decisions needed to be made. Ferrari called a meeting on May 25. His inner circle gathered around, although Surtees was not invited. The boss asked for everyone's opinion regarding the Surtees situation and the staff was divided. Some were for keeping him, some for sacking him.

"At the moment," weighed in Gozzi, "we can't lose our top driver. Bandini isn't a number one. When I spoke to Andretti yesterday he said that he could not join us for at least another year."

Dragoni saw an opportunity and he jumped on it. "As well as going into the F1 workshop," he said, "Surtees also had a look at the sport 330 P3 and—surprise—in London Eric Broadley is building a Lola sport that's absolutely identical."

The team manager had made a serious accusation—that Surtees had stolen design secrets from Ferrari and brought them home to England. But was it true? Surtees had a relationship with Lola and Eric Broadley, it was no secret. He had been in a Lola when he crashed seven months earlier in Canada. Team espionage was a fireable offense, period, and Surtees wasn't present to defend himself. ("This is the most stupid thing of course that ever came out," Surtees said in retrospect. "Anybody with the *slightest* bit of knowledge would look at [the Ferrari 330 P3 and the Lola] and see there wasn't the *slightest* resemblance.")

Ferrari thanked everyone and called an end to the meeting. He needed time to think. All he could know was what he was told by others, since he didn't attend races and was preoccupied by the Fiat negotiations. He was a highly diplomatic man. It's unlikely that he believed Surtees was guilty of anything but impetuousness and an absolute *need* to win. Nevertheless, he made his decision. He pulled Gozzi aside and assigned him a special mission. Gozzi was to go to the Belgian Grand Prix on June 12, the weekend before Le Mans. He was to appear in the press office and announce the immediate firing of John Surtees.

"You go and make the announcement at the end of the race," Ferrari said. "Nothing else."

Gozzi set out with two journalists from the magazine *Autosprint*, bound for Belgium. It was an exhausting 18-hour drive, the journalists interrogating Gozzi the whole way. They could smell the scoop but Ferrari's lieutenant wasn't talking. They'd have their story soon enough.

Spa-Francorchamps was possibly the most beautiful racing circuit in the world. It wound through wooded hills and livestock farms, through the Ardennes forest, on public backcountry roads where farm tractors normally trudged at creeping speeds. It was murderously fast, with looping bends and severe elevation changes.

From the beginning, Surtees put on a spectacular display of skill. He qualified first and when the race started, the rain came. In the showers, he tucked in close in second place behind a Cooper-Maserati, letting the Austrian Jochen Rindt's deeply grooved Dunlop tires cut a path through the puddled pavement. This strategy relied on Rindt not making a mistake, as through the whitewater kicked up by those Dunlops, Surtees was driving, at times, completely blind through fast bends and elevation changes at furious speeds.

On the penultimate lap he passed Rindt, moving into first place. A camera helicopter followed him around the 8.7-mile circuit. The director John Frankenheimer was filming *Grand Prix;* the cameras captured Surtees's brilliant performance in the #6 Ferrari as he lapped car after car in the rain. In a few months' time Americans would sit in their theater seats and watch in awe as the red missile fired across the big screen.

In only his second F1 race after his critical injuries, Surtees won the Belgian Grand Prix. He stood on the podium and waved at the crowd, holding a bouquet of flowers so vast, it engulfed him. He got no congratulations in the Ferrari pit. When he saw Dragoni, the team manager criticized his performance, claiming Surtees trailed in second place for too long.

Surtees was incredulous. "Look," he said, "when you tell me how to drive my races, that will be the day. Winning is the only thing that matters."

Gozzi ran to a telephone to report back to Ferrari. As he placed the call, the Italian national anthem was blaring in the background. "He won. What should I do?" He heard a moment of silence on the line.

"Suspend it," Ferrari said. "And come back immediately."

Le Mans was six days away.

THE FLAG DROPS

JUNE 1966

> This racetrack is a cornfield airstrip in the jet age. It was built 50 years ago for cars that went 65 mph. Tomorrow 55 race cars—some of them capable of 225 mph on the straightaway and all of them over the 130 mph class—will get off at 10 A.M. (Detroit time) and it will be a miracle if nobody gets killed. Nobody is fearless. Some of these drivers are scared stiff.
>
> —*Detroit News*, June 17, 1966

IN LEO BEEBE'S OFFICE, the phone wouldn't stop ringing and the paperwork was piled high. Days before departing for France, Beebe was struck by a string of setbacks. Eight Fords would be entered at Le Mans, which meant the company was contracting at least sixteen of the top drivers in the world. In a single week in early June, a handful of crashes on racetracks cut into the available talent. The company found itself short on drivers.

June 4: Speeding into a turn at the Milwaukee Fairgrounds, A. J. Foyt lost a front wheel. He hit a wall and his car exploded. "I knew I had to get out or just fry," he said from his hospital bed, his hands, neck, and face scorched and bandaged. "I gritted my teeth and put my hands into the burning fuel to raise myself out." Foyt was under contract to race a Ford at Le Mans, but he wasn't going to make the trip.

That same day: Ken Miles's teammate Lloyd Ruby went down in

a plane crash after taking off from Indianapolis Motor Speedway airport, suffering a fractured spine.

June 12: At the Belgian Grand Prix, Jackie Stewart—also contracted to race for Ford at Le Mans—lost control on wet pavement and launched through a wall of hay bales into a nine-foot ditch. He survived a harrowing ordeal, pinned beneath a smoldering car while leaking fuel soaked into his coveralls. He suffered a fractured shoulder and ribs.

Phil Hill had a falling out with Ford management. Hill signed with another team that posed a major threat: the Chaparral of Texan Jim Hall, with its big Chevrolet engine. Hall had rebuilt the car to meet all FIA regulations and had hired the most accomplished Le Mans pilot of them all.

Leo Beebe needed drivers, but who was capable of taking the wheel of a 225-mph machine? You couldn't put that kind of power in just anyone's hands. "The Flying Scot" Jimmy Clark—two-time F1 champ and reigning king of Indy—refused to race at Le Mans. It was simply too dangerous. So, too, did Fred Lorenzen (the first NASCAR driver to earn more than $100,000 in a season, in 1963), for the same reason. At the last minute, Ford reached out to the new kid.

Not since his family left a refugee camp in Italy had Mario Andretti stepped foot in Europe. Ford Motor Company offered a big payday. Andretti saw a chance to get some experience across the Atlantic, without having to make the kind of commitment that the Ferrari offer would've required. And the money was too good to turn down. Andretti packed his bags.

The weekend of June 12, racing teams from all over Europe and America mobilized. At Shelby American in Los Angeles, Holman Moody in Charlotte, and Dearborn, the rush to move personnel and equipment overseas got under way. The logistics were staggering. Eight 2,400-pound cars, four spare engines, more than 25 tons of tires and parts, and a 40-foot tractor trailer turned into a rolling machine shop were stuffed into the bloated bellies of jetliners. The staff totaled more than 100 men. In Dearborn, Beebe's team arranged for Ford's own medical unit. "We have our own medical tent, our own doctor, and a helicopter standing by to evacuate anybody who gets busted up," he told one reporter.

As usual, Shelby enjoyed his share of the prerace spotlight. He was coach and captain of Henry II's army, a Pied Piper of hotrodders, a tough Texan with a sissy's name. Shelby had gotten word that Andretti had joined the Ford effort. The young sensation was contracted to Firestone, which meant he had no choice but to race for the Holman Moody Ford team. One more thing for Shelby to worry about.

"Win or lose," he said on his way out of Los Angeles, "I'm flying home on Monday and I'm going to go hide for awhile where no Ford man or Frenchman can find me."

In Paris, Henry II arrived by jet with his wife and son Edsel II. He was to spend the whole week in France. On Monday, June 13, he'd meet with Prime Minister Georges Pompidou to discuss business and the European economy. On Wednesday, his appointments included France's Finance Minister Michel Debré. And on Saturday, he would arrive at the racetrack at Le Mans as Honorary Grand Marshall. His arrival was treated like the second coming of Henry II of the House of Plantagenet, the English king who ruled much of France during the twelfth century.

Henry II was about to pump another $100 million into overseas operations by the end of 1966, and the French leaders were vying for some of those robust American dollars. They had a new French Ford factory in mind. Ford's growth in Europe was rising 5 percent to 7 percent a year, double growth in the United States. Although in America Ford Motor Company was second to General Motors, in Europe, among American manufacturers, Ford was number one.

The day Henry II met with Prime Minister Pompidou, Ralph Nader appeared in Europe for the first time to spread his message. Was the timing coincidence? Lectures in Great Britain drew crowds and banner headlines across the continent. Nader's face was the face of controversy, his voice the voice of reform. "The road death and injury rate in Europe has reached tragic proportions," Nader preached. "They have not done anything about safety since Magna Carta."

In the city of Le Mans, hotels filled with the world's elite racing drivers, mechanics, and engineers, the price of each room jacked up to double or triple the normal rate. Journalists and television crews arrived, along with spark plug sales reps, tire company executives, and anyone else who stood to make a buck. Sleepy bars in town cranked their jukeboxes and the streets grew loud with late-night, drunken foot traffic. Ford men filled two hotels downtown. Ken Miles was spotted outside the Automobile Club de l'Ouest's official inspection tent. Standing shirtless in light blue chinos and shades, he shook hands and signed autographs. Andretti checked in to the Hôtel Moderne. He got off to a poor start in France; he was in a dark mood, brooding over bad news. The night before the first practice day, he sat down to dinner with his wife at the hotel.

"Did you hear what happened at Reading?" he asked. She hadn't heard. "Jud Larson was killed and so was Red Riegel." Larson and Riegel were two dirt-track racers who'd schooled Andretti in his early days in Pennsylvania.

When asked about Le Mans, Andretti said that he'd do his best. Sports cars weren't his specialty but "money talks," he said, sarcastically. "We're all here to make money. What else? That's what Jud Larson was doing at Reading, pushing to make a buck. But he pushed too hard."

Phil Hill showed up with the Chaparral team. If the Chevy-engined Chaparral took the checkered flag as Henry Ford II sat in the crowd watching, it would go down as one of the most humiliating moments in racing lore, and many in the know were putting their money on Hill.

The Ford team set up operations in the old Peugeot garage not far from the track. The garage's floor space was the size of a football field, but, by Wednesday, June 15, the team was already running out of room. Bruce McLaren walked into the garage with his teammate Chris Amon. McLaren saw eight cars painted in various colors. Dozens of technicians worked over the fleet in spotless Ford uniforms.

McLaren had been the first driver hired by Ford to develop a Le Mans car, back in 1963. He couldn't believe how massive the operation had become. But then, that was racing in 1966. Money was pouring in, advancements in technology were throttling forward, and speeds kept rising. McLaren had himself cemented his reputation during the previous few years as one of the most innovative engineers and racers in the world. He'd launched his own Formula One team just a month earlier, fielding cars under the McLaren marque. It seemed the natural course of things since his childhood days, when he raced around the Wilson Home for Crippled Children on his Bradshaw Frame. McLaren might have known he would someday die at the wheel. He'd just published an autobiography in which he'd written his own epitaph: "To do something well is so worthwhile that to die trying to do it better cannot be foolhardy. It would be a waste of life to do nothing with one's ability, for I feel that life is measured in achievement, not in years alone."

Shelby believed McLaren's expertise was so valuable, he'd negotiated to have him on his team even though McLaren was contracted to Firestone and not Goodyear. No one knew at the time that this little footnote in the story of the Tire Wars was to play a strange and significant role in the outcome of the race.

With his teammate Amon, McLaren found his car. They were delighted by the sight of their Ford. Both he and Amon were from New Zealand, and their car was painted black with silver racing stripes—kiwi sporting colors. The number two was painted on the nose and doors. Neither driver had ever seen a car better prepared for competition than this one. Many critical pieces had been X-rayed. There was no need for frantic, last-minute work; everything was ready to go.

Thus far, all of McLaren's races for Ford had ended in disappointment. Months earlier at the 24 Hours of Daytona, he and Amon had teamed and had taken it easy on their Ford fearing it would break—as it always had before. They'd finished in fifth place, miles behind the leaders. They'd "pussyfooted," in Amon's words, because they didn't *believe* in the car. After that race, McLaren had told his friends that Ford was going to win Le Mans in 1966. Now here he was in France, about to start the biggest race of his life. There'd be no pussyfooting, he told Amon.

"We're not screwing around like that again," McLaren said. "We're going to *go* for it!"

Surtees came to Le Mans directly from Belgium following his victory at Spa-Francorchamps. He knew he and his team were underdogs. After six straight Le Mans victories, Ferrari was now expected to lose. Surtees had a strategy in mind. When the team had gathered—roughly two dozen men in comparison to the Ford army—he stated his case.

"The only way we're going to beat these cars is by playing tortoise and hare," Surtees said. "Mechanically, our cars are well engineered and they can be driven 99 percent. The thing is, you got racers at Ford. *Real racers.* Now I reckon we can get away with a bit of luck by driving one of our cars at 100 percent. The moment the flag drops, we have to go—*bang!* We can run our other cars with a little more safety. I cannot see some of those drivers resisting, not joining the race. I can see their discipline easily go right out the window. We can win the race that way!"

It was a plan. One Ferrari would charge out front at the start and try to bait the Ford drivers into moving too fast too early and breaking their cars. Time and again, the Fords had proven fragile. The Americans understood *power,* but were they the best engineers? The most disciplined drivers? Surtees was the only man on the team quick enough to make this strategy work. He'd be the man to set the pace.

The afternoon of Wednesday, June 15, mechanics from all the team garages wheeled cars into the pits and onto the circuit for the first practice session. Surtees was on the track, as were Miles, Gurney, McLaren. On the fastest straightaway in racing, the world's most skilled drivers busted the 200 mph mark again and again, the thin strip of pavement lined on either side by sturdy pine trees. By the time the teams left for their hotels after the first day, the sun had gone down and the lap record was history. Dan Gurney had set a new mark in a Ford—3 minutes 33.3 seconds. Surtees was three seconds slower, but still warming up.

Surtees arrived in the Ferrari garage the following day prepared to beat the Ford's time on the second of three practice days. He found his 330 P3. The names of the drivers for each team car were painted in white on the red body. When Surtees saw the names on his car, he knew something was wrong. He was supposed to be teamed with another Englishman named Mike Parkes, but there was a third name painted on his Ferrari, Ludovico Scarfiotti, a fast Italian who happened to be the nephew of Fiat boss Gianni Agnelli. Because of his health, Surtees had agreed to allow Dragoni to assign Scarfiotti as a reserve driver to his car. But Surtees was feeling strong, and Scarfiotti was reserve, on the bench. The Italian's name wasn't supposed to be painted on the car. Surtees knew something was up, and Dragoni was behind it. He went in search of his nemesis.

At Le Mans, the world caved in on Surtees. The near-death accident, the stress his return to racing had placed on his marriage, the backstabbing and cutthroat politics. Even worse: Dragoni had put up a fuss about Surtees's wife Pat sitting in the pit. Pat Surtees always charted her husband's lap times. There were rumors that one of the Ferrari engineers was making passes at her, and Dragoni said it was distracting. Making passes at his wife while he was on the track risking his life! Was it true? Who knew? All of Surtees's crackling rage funneled into this one moment in time.

He found Dragoni.

"Why is Scarfiotti's name painted on my car?"

"Perhaps you are not fit enough," Dragoni answered. Then he added, "John, you are not going to start. Scarfiotti will start."

"Why?" Surtees said, exasperated. "That's against all the philosophy! That's against what we've said is necessary to win!"

"Mr. Agnelli is here," Dragoni said. "He's here for the start. I want Mr. Agnelli to see his nephew Scarfiotti."

"Ludovico's not going to be able to mix it up with the Fords in the early part of the race! We have a strategy in place!"

The strategy had apparently changed. "Besides," Dragoni said, "we thought if we let Ludovico start it would give you an easier time. Perhaps you might be tired."

Surtees realized what was behind all of this. As head of Fiat, Agnelli was the man that Enzo Ferrari was currently negotiating with for a strategic partnership, and diplomatically it would make nice if Agnelli saw his nephew start the race. Surtees had gotten caught up in the Fiat affair. And Dragoni, Surtees believed, had his own agenda. Surtees and his teammate Mike Parkes were both Englishmen. Dragoni wanted an *Italian* on their team so the Brits wouldn't get all the credit if they won.

Surtees dug in. A shouting match erupted, and it ended with Dragoni laying down an ultimatum. If Surtees didn't like the scenario, he could walk.

"I'm the fastest man on this team," Surtees responded. "Whatever is behind this, it's not in the best interest of the team. The main job is beating Ford and winning this race. I'm tired of constantly being sabotaged in my efforts to win by decisions that make no sense. Therefore, you'd best count me out!"

He stormed off. Within minutes, word began to spread. An extraordinary new development reached the pressroom and reporters went in search of the facts. This was no minor affair. Surtees's refusal to compete at Le Mans was like Hank Aaron refusing to suit up for Game Seven of the World Series. The driver sent a telex to Maranello pleading with Ferrari to do something about the Dragoni situation—to no avail. Ferrari did eventually comment publicly: "If Dragoni has decided on the substitution, then he must have good reason for doing so."

It occurred to Surtees that all Mr. Ferrari ever knew to be truth was funneled to him by a variety of pawns in his Medici-esque empire. Who knew what Dragoni had told Ferrari in the past, and what he'd say about this morning's row? Surtees had to get his own story out there immediately. When the reporters arrived, he was ready to talk. He spoke to them all: the English reporters, Italians, Americans, even ABC's television cameras, filming footage to fill out the weekend's live broadcast.

"Dragoni and I have never agreed," Surtees said. "The big problem is he is an Italian and I have not got the right nationality. Things came to a head this morning. I was asked by Dragoni if I agreed to the Italian Ludovico Scarfiotti becoming the reserve driver to the three Ferrari works cars. Mike Parkes and I agreed. I was astonished when I arrived to find three names were painted on the one car which Mike and I were to have driven. When I tackled Dragoni he said something about me not being fit enough to finish the race. I told him not to make lame excuses. His reply was, 'Take it or leave it.' I chose to leave it.

"We came here to beat the Fords," Surtees continued. "When things are going on in the pits which tend to stir up trouble among the drivers, this is not my idea of a team effort."

Surtees knew he had to go see Mr. Ferrari in person and make his side of the story known before it was too late. He knew this meeting had the potential to be the most explosive of his life.

By Friday night, the qualifying times were set. Dan Gurney was fastest in the #3 Ford: 3 minutes 30.6 seconds. Ken Miles was second, 1.1 seconds behind. The quickest Ferrari qualified fourth at 3:33.

The night before the race, drivers struggled to sleep. Some picked anxiously at late-night dinners, while others lost themselves in the eyes of some beautiful conquest. In a castle called Château d'Artigny not far from the track, Ford executives gathered for a VIP cocktail party. They walked wide-eyed around this massive Renaissance building, swallowing hors d'oeuvres with glasses of French wine. How different this was from the usual Dearborn fete. None of the waiters spoke English, and though they liked American money, they didn't like Americans.

At one point, Roy Lunn ventured to the bathroom. Though the building was majestic and expensive, the toilet had a trough-like urinal that seemed centuries old. He stood thinking about how this whole European campaign had unfolded. Three years earlier, he had been at the Ferrari factory with Mr. Ferrari on a fact-finding mission. He could recall the Dearborn meeting after the Ford–Ferrari deal had gone sour, when he outlined his proposal before Lee Iacocca and Don Frey to build a mid-engine racer that could beat Ferrari at Le Mans. Now Lunn was relieving himself in this castle southeast of Paris the night before the most anticipated race in Ford Motor Company's history.

He heard the bathroom door open. Out of the corner of his eye, he saw a large man amble up next to him. It was Henry Ford II. They stood next to each other.

"Well, Roy," Henry II said, "do you think we're going to win?"

Lunn searched for words. He was caught off guard, though pleased Mr. Ford had remembered his name. "I hope so," he said. "I think we've got a great shot."

Around the globe, headlines alerted millions that international racing's ultimate showdown had arrived. Stories appeared in Paris's *Le Monde,* Milan's *Corriere della Sera,* the *New York Times,* and the *Los Angeles Times*. The question would finally be answered: Which was faster, Europe or America?

Ford had entered eight prototypes, all Mk IIs with 427-cubic-inch engines. The final lineup:

- » Ken Miles and Dennis Hulme (entered by Shelby American)
- » Bruce McLaren and Chris Amon (Shelby American)
- » Dan Gurney and Jerry Grant (Shelby American)
- » Paul Hawkins and Mark Donohue (Holman Moody)
- » Ronnie Bucknum and Dick Hutcherson (Holman Moody)
- » Lucien Bianchi and Mario Andretti (Holman Moody)

» Graham Hill and Brian Muir (Alan Mann Racing)
» John Whitmore and Frank Gardner (Alan Mann Racing)

Ferrari had prepared seven prototypes officially entered by various teams, though all were prepared at the Maranello factory:

» Mike Parkes and Ludovico Scarfiotti (Factory-entered 330 P3)
» Lorenzo Bandini and Jean Guichet (Factory 330 P3)
» Pedro Rodriguez and Richie Ginther (Luigi Chinetti 330 P3)
» Masten Gregory and Bob Bondurant (Chinetti 365 P2)
» Richard Attwood and David Piper (Maranello Concessionaires* 365 P2)
» "Beurlys" and Pierre Dumay (Ecurie Francorchamps† 365 P2)
» Willy Mairesse and Hans Müller (Scuderia Filipinetti‡ 365 P2)

A supporting cast of privateers had entered five more Ford GT40s and four Ferrari GT cars. Ferrari also entered three Dinos, set to battle in their own 2-liter race against the Porsches. In total, fourteen Ferraris and thirteen Fords were entered.

Race day morning arrived with the threat of rain. By lunchtime, hundreds of thousands had descended upon the grounds. All morning, the sky darkened. In some of the pits, the crews were busy changing from dry to wet tires, expecting the worst. Was it going to rain or not? The choice was critical. In the Ferrari pit, all talk was of the Surtees-Dragoni disaster. Dragoni answered to the Italian reporters.

"Let's drop the subject that I favor Italian drivers," he said. "Here I work only in Ferrari's interests."

Where did Enzo Ferrari stand on the matter?

"I didn't speak with him because the telephones weren't working well," Dragoni answered.

Surtees would not be able to meet with Ferrari until after the

* Maranello Concessionaires was a British-based professional Ferrari racing team.
† Ecurie Francorchamps was a Belgian-based professional Ferrari racing team.
‡ Scuderia Filipinetti was a Swiss-based professional Ferrari racing team.

race. Back at his hotel, he received a strange note. It was from Carroll Shelby, asking if maybe Surtees wouldn't mind a drive in a nice little Ford racing coupe. Certainly Shelby could clear a spot on his team.

Surtees packed his bags and prepared to head south to Italy.

Secluded from the crowds, Ford's sixteen drivers reported to their final briefing exactly two hours before the start. They were dressed in fireproof coveralls and asbestos-bottomed racing shoes. Shelby and Beebe stood by. Fingertips tapped nervously on the tops of helmets. Some of these drivers were brought in at the last minute to fill out the ranks, having never competed at Le Mans before.

A project manager under Beebe named John Cowley did the talking. The executives understood there was a tremendous rivalry between Shelby's drivers and Holman Moody's. There was rivalry between Shelby's drivers themselves; Miles (in the #1 Ford), McLaren (#2), and Gurney (#3) were at Le Mans to win. For McLaren, a Le Mans victory would make for incredible publicity for his Formula One team, which was two races old. Many believed Gurney was, all around, the best in the world. Miles had rubbed some Ford men the wrong way. Sure, he'd done most of the development work on the car. But he had a reputation for having too much race in him. That stunt at Sebring, battling with his teammate Gurney, had landed Miles in hot water. Was he a team player?

"Miles would race his grandmother to the breakfast table," Ford racing executive Jacque Passino said. It wasn't a compliment.

The company men had to drill into the drivers' heads that this was a team effort. *There was to be no interteam competition.*

"I appreciate that you all have been in racing a lot longer than I have," Cowley said. "We want you to lap at a pace consistent with both finishing the race and breaking the Ferraris." He then assigned lap speeds for the start of the race. Gurney, who'd won the pole, was given the fastest time: 3:37–38. Miles was to lap two seconds slower, and McLaren two seconds slower than Miles. The effort would be highly coordinated and disciplined. Under all circumstances, drivers were to follow pit signals and team rules.

"For maximum durability, do not exceed 6,200 rpm," Cowley said. "Driver changes will be at every other fuel stop." While it was custom for drivers to wait until they reached the Mulsanne Straight to buckle their belts, the starters were told to strap in when they jumped in their cars, *before* pulling off the line. And all attempts would be made to conserve the brakes. Were there any questions?

"Okay," the Ford man said. "Let's get it done."

As 4:00 P.M. neared, Le Mans's guest Grand Marshal arrived. The Deuce's helicopter touched down in the airfield behind the grandstands. He was accompanied by his son Edsel II and his wife. By this time an all-time high 350,000 had crowded onto the grounds around the circuit. When Henry II walked onto the pit lane among the crowd, he was met by a host of luminaries. Iron-gray hair slicked back, his prodigious girth cut a path through the crowd. His wife had bet $1,000 that Ferrari would win. "After all," she laughed, "I am Italian!"

Reporters peppered Mr. Ford with questions. Why was success at Le Mans so important?

"Ford is an international company," Henry II said, "with branches all over the free world. We feel a good showing by our products at Le Mans will reflect favorably on us in the countries where we do business."

What was he hoping to see?

"Aside from victory, I hope to see interesting competition. And I especially hope to see a safe race, without accidents."

Henry II stopped in front of one of the Ferraris. He stood there holding his hands behind his back, staring down at this Italian automobile. It was indeed a thing of great beauty and power, its curves organic and undeniably sexual. From out of the crowd, Leo Beebe's tall frame appeared. Henry II looked up into Beebe's eyes and shook his hand. The two men had known each other for more than two decades, since their days in the navy. Henry II reached into his pocket, pulled out a card, and passed it to Beebe. Then he continued onward.

Beebe looked down at the card. It read, "You better win, HF II."

He put the card in his wallet. It remained there for the rest of his life.

At 3:59 P.M., the crowd grew quiet. The grandstands looked like they could sink into the earth with the sheer weight of bodies. Atop the grandstands, the flags of countless nations waved in the breeze. Thunderclouds pressed from above.

Every eye was on Henry Ford II. As Honorary Grand Marshall, he had the privilege of starting the race, and he stood on the pavement holding a French flag high. Twenty yards from him, Ken Miles stood across from his #1 Ford, which was painted orange and light blue with white racing stripes streaking down the nose. His helmet was black and so were his shoes. Miles had traveled a long, twisty road to get to this place in time. The pavement beneath his feet was drenched in glory, courage, honor, and blood.

At precisely 4:00 P.M., the Deuce dropped the flag. Then he hustled off that pavement as fast as his legs could carry him.

22

LE MANS – RECORD PACE

JUNE 18, 1966

MILES KNEW RIGHT AWAY he was in trouble. When he jumped into the cockpit and slammed the door shut, he smacked his helmet on the lip of the door. He took his time and clicked on his seatbelt. Then he hit the ignition, stood on the pedal, and up-shifted through the gears, firing off the line with the rest of the pack. As he moved under the Dunlop Bridge, planting all of the 427's torque, he realized he was going to have to pit after a single lap.

He'd bent the door by slamming it on his head.

Carefully he maneuvered the twists and the back straight. One wrong move and the 24-hour race could turn into a two-minute-long calamity. When Miles pulled into the pit after that first lap, Shelby feared the worst. Mr. Ford was watching; the Deuce was right there in the pit. The French announcer shouted over the loudspeakers: the #1 Ford had come in for repairs. Miles opened his door and screamed at his crew chief. They moved quickly, under Shelby's watch. By the time the door was fixed, Miles was already behind. He accelerated onto the track and began to weave through the heavy iron and the smaller cars.

The front-runners fell into order and Dan Gurney set the pace as he was instructed, leading the race in his bright red #3 Ford. Miles chased in pursuit. A drizzle began to fall and he clicked on his windshield wiper. He saw a sea of umbrellas open around the track. As the pack stretched out in single file—carving the black stripe into the pavement that was the racing line—the real competition began. Miles throttled through the grandstands and the announcer heaved over the loudspeakers: the #1 Ford had shattered the lap record at 3 minutes 34.3 seconds. By the end of the first hour, Miles had moved into third place. Fords were running one through four, their throaty Detroit engines loud and proud. There was no finesse to the mighty V8's song. It was bone-hard force, a great metal fist flying down the road. The announcer came again: Miles had lowered the lap record to 3:31.9, an average of 142.01 mph.

When the fastest Fords came in for the first scheduled pit stops, Shelby watched his men work with military precision. He saw careful planning and perfect execution. Miles pulled the #1 Ford in at 5:26 P.M. His crew checked his brakes, Goodyears, oil, and fueled the tank with 33.5 gallons. He pulled out of the pits in a flash. Gurney followed at 5:30 in the #3 Ford. When McLaren came in at 5:33, he stepped out of his car and complained about his tires. He could feel the track tearing up the rubber. His crew had a look; the Firestones were chunking. Nobody on Goodyears was having any trouble. McLaren's teammate Amon stepped in and they had a conference with Shelby.

"It's up to you guys," Shelby said. "You're contracted to Firestone but we've got plenty of Goodyears sitting here."

Soon McLaren was back on the track on a set of Goodyears. He'd lost time, vital minutes that would prove costly at 4:00 P.M. the next day.

Speeds continued to rise. To see these cars curving through the Esses was to witness the pinnacle of engineering and human courage. The announcer piped over the loudspeakers: it was Gurney in the #3 Ford, followed by Miles in the #1 Ford. The Ferraris stalked less than a minute behind, awaiting the right moment to strike. The Shelby American Fords were well faster than the Holman Moody cars, which were trailing back in the pack. Andretti was nowhere near the leaders. Miles lowered the lap record again, then Gurney stole the record with an incredible 3:30.6. In the press box, ABC's Charlie Brockman shouted into his microphone.

"We are witnessing the most tremendous 24-hour racing ever put on here at Le Mans, as it has been a battle from the start among four or five cars at the most tremendous pace of all time. Working with us this evening is a man who's won here three times and is a competitor in this one, Phil Hill of Santa Monica, California, who will be driving the #9 Chaparral, right now lying in 10th position. Phil, the talk here has been on the pace of this race."

"It's a terrific pace," Hill said. "Before the thing's over, we're going to see an awful lot of cars not around anymore."

"The lap record, which you set here a year ago at 3:37.5, has been broken. It's now held by Dan Gurney at 3:30.6, an average speed of 142.89 mph. That's fantastic."

"It sure is."

At 6:47 P.M., McLaren pulled into the pit for the first scheduled driver change. He was loving the car. "They were *so* fast, those 7-liter Mk IIs," he later recalled. "We were doing 220-something mph down the Mulsanne Straight." He jumped out to confer with his teammate Chris Amon as the mechanics checked the brakes and tires and pumped in 32.3 gallons of Super Shell.

"Is it okay?" Amon asked.

"Yeah, fine," McLaren said.

Amon jumped in and steered the car back onto the track.

Miles pulled in one minute after McLaren and handed the car over to his teammate Denny Hulme, a New Zealand-bred Formula One racer who'd replaced injured Lloyd Ruby. Hulme was celebrating his thirtieth birthday in style that day—by racing at Le Mans. The crew popped the #1 Ford's engine compartment, checked the brakes, and fueled 33.4 gallons.

Shelby checked the lap chart. Miles had driven his first stint averaging 3:37.2 laps—three seconds faster than Gurney, who was in the lead. If Miles hadn't smacked his helmet on the door at the start, he'd be well in front.

The crowds in the grandstands saw a burst of black smoke in the distance climbing out of the trees above the Mulsanne Straight. Someone had crashed hard. But who? Wives and girlfriends grew nauseated. The camera helicopter hustled over for a bird's eye. ABC's Charlie Brockman:

> There you can see [from the helicopter's camera] the fire that blazed up when two cars got together on the Mulsanne Straight. There's an eerie feeling to be over here on the pit straight and not know who is involved. Immediately the question starts going around: Who is it? Who is it? The fire has crept into the woods. The trees are burning, soaked in spilled gas and oil.

As drivers maneuvered the *Tertre Rouge* corner onto the long straight, they saw a flagman waving yellow, but they didn't need a flagman to know there was danger. Smoke clouded the roadway. As they passed the wreck, they saw the flames on their left spread out over two dozen yards, with no rescue squads in sight. Two wrecked cars smoldered. A quick glance moving by at speed and anyone would assume that funerals were in order. The more experienced drivers hit the gas hard, knowing the others were letting up. Soon the facts were reported over the loudspeakers: two French drivers were en route by helicopter to a nearby hospital. One had minor injuries, the other had two broken arms.

Nearing 10:00 P.M.—sunset at that latitude in Europe—the clouds on the western horizon turned a shade of blazing orange. When Henry II's helicopter lifted off, headed 50 miles away to Château d'Artigny for a scotch or two and a night's rest, Miles's #1 Ford was leading the race (with Hulme now at the wheel). But the Ferraris were still chasing fast, and a fleet of silver Porsches were in the running, too.

Through the opening hours, the Ferraris had remained within striking distance. When Hulme brought the #1 Ford into the pit to hand back over to Miles, the crew went to work changing the front and rear brake pads and the tires. The red cars stayed on the track. The Fords possessed superior speed, but the Ferraris were more nimble and efficient. While Miles's car was in the pit being worked over, the Ferraris made their move, pulling into the lead for the first time.

Reporters in the press box working to make deadline for the next day's paper hammered at their typewriters. Six hours into the race, it was one-two for Ferrari.

In darkness the rain came, pounding the crowds and speeding cars. Spectators ran for shelter and headlamps charged through the murky night. Car after car veered off into the pit lane to change to wet tires.

As the pace slowed considerably, and the temperature dropped, Miles started his second shift and began to speed through lightning-fast laps. He knew this car so well, he could feel the rubber's grip on the wet pavement. The driving was safest in the racing line, where tires had left a trail of rubber gunk. On either side was where the pavement was slickest. In order to pass, one had to veer at times from the safest line into that slippery tarmac, made even more treacherous by the cooling temperature. Miles overtook car after car on the outer edges, moving 50, 60, sometimes 70 mph faster. He moved back into first place and began to draw out a lead.

Shelby and Ford management retreated to a dimly lit corridor behind the pits to shield themselves from the squall. The alley led under the grandstands to the paddock on the other side. It was long and lit with fluorescent lights, with cinder block walls and a cement floor, pipes running along the ceiling. Spectators forbidden. Until there was some exigency, all these men could do was wait. They paced the dirty floor "like expectant fathers in a hospital waiting room," as one man present described the scene. The sound of engines echoed down the corridor.

Someone came down the hall with an updated lap chart. Miles was lapping at 3:39. The signaling pit was giving him the EZ sign but he wasn't slowing down. He was moving quicker than he was ordered to go in the fast opening laps. And now it was pitch black out there. And pouring.

"The old man is really running in that rain," Shelby said of Miles.

The pacing continued.

23

THE MOST CONTROVERSIAL FINISH IN LE MANS HISTORY

JUNE 19, 1966

THE MIDNIGHT HOURS proved the most vital and frantic at Le Mans in 1966. The murderous racetrack took its toll, tearing up machinery and men.

Andretti's Ford blew a head gasket after 97 laps. His adventure at Le Mans was over.

Phil Hill's car lost its headlights—dead battery. The Chaparral was wheeled out of the pits for good.

Ludovico Scarfiotti was moving at high speed into the loopy Esses in darkness in the #20 Ferrari. Screaming into a blind turn, Scarfiotti found a wreck right on the other side. Two banged up cars were just sitting there. He had no choice. He pounded into the obstruction with a loud wallop, littering the pavement with pieces of his car. Scarfiotti was severely shaken, both by the accident and by the fear of Mr. Ferrari's wrath. The Italian fans would hang poor Scarfiotti out to dry. He'd taken the place of *Il Grande John* in the lead prototype, and had failed to make it through the night. He was in the hands of medics now.

Just after 3:00 A.M., word went out over the loudspeakers that the most threatening Ferrari, the 330 P3 driven by Pedro Rodriguez and Richie Ginther, had broken its gearbox. In a matter of three short hours after midnight, fate placed the race in the hands of Carroll Shelby and his team. His three entries were running one, two, three. All the front-line Ferraris were out. The closest Ferrari was in fifth place, dozens of miles behind. Only one Holman Moody Ford was still running and it was nowhere near the front.

Looping around the circuit in the darkness, the Shelby American drivers got pit signals to slow down and lap no faster than four minutes. Miles's #1 car was in the lead, Gurney's #3 car second, McLaren's #2 car third. The drivers had their orders: No interteam racing. The only things that could change the finishing order were mechanical problems and pit stops.

At 4:10 A.M., minutes before the first hint of sunrise, Miles pulled into the pit. And he stayed there. The announcer began to scream over the loudspeakers. The leading car seemed to be in trouble. The pit crew was working over the car but no one outside the pit had any clue as to the problem.

It was the brakes. Miles had worked those disc brakes hard. His pit crew had practiced for days preparing for this moment. They went to work changing the front brake rotors, using the quick-change system that Shelby's chief engineer Phil Remington had designed. In minutes they finished a job that would've taken a small-town team of mechanics several hours. The French officials started bickering. Was this legal? They'd never seen rotors changed so fast. Miles's crew performed the same procedure on the rear brake rotors at 7:34 A.M. Crafty pit work saved the race for Miles. The sun was up, the rain had subsided, and the #1 Ford was hammering down the road.

Miles was now in second place behind Gurney's car, but a few minutes after 9:00 A.M., the #3 Ford pulled into the pit with a radiator leak. The water temperature gauge was pinned. Gurney couldn't believe it. First Sebring and now Le Mans. His luck couldn't get any more rotten. But some in the pit didn't believe he had bad luck. They said he'd pushed his car too hard because every time he had the lead, Ken Miles was right behind him, nudging.

"There was some talk in the pits that Ford thought that Ken did not follow team orders and pushed Gurney to the point of breaking," Miles's crew chief Charlie Agapiou later said. "That was absolute bullshit. Ken followed his directions to the letter. When he pitted on lap one to fix the door, he lost several places. After returning to the race, he had to go like hell in order to get back in second place behind Gurney. That's where he was told to be."

At 11:00 A.M., Henry II's helicopter landed at the airfield behind the grandstands. He and his family made their way to the Ford pits, where they were filled in on the latest developments. It had been a long, brutal night. With five hours to go, Fords were running one, two, three, and four. Miles's #1 car was leading, followed by McLaren's #2 car. It appeared Mrs. Ford was going to lose her $1,000 bet.

Shelby watched the cars speed by, his eyes bloodshot and irritated. The sun peeked from behind clouds and disappeared again. The hands on the Dutray clock moved slowly. Around and around the leaders drove, the cars now battered and bruised, coated in a layer of soot. They were so filthy, their racing numbers were partly obscured. With roughly two hours left, Leo Beebe, Don Frey, and other members of the Le Mans Committee crowded into the pit. They were going to win Mr. Ford's war against Enzo Ferrari, that much was clear. They were going to get to keep their jobs. But still there was an air of sobriety, a sense that the unexpected was still imminent. Conversation turned toward the finish. The suits asked Shelby how he thought this story should end.

"Well hell," Shelby said as cars darted past a few yards away. "Ken's been leading for all these hours. He should win the race." Shelby looked at Beebe. "What do you think ought to happen Leo?"

Beebe thought for a moment. "I don't know," he answered. "I'd kind of like to see all three of them cross the finish line together."

An interesting proposition, Shelby thought. If the first-, second-, and third-place cars crossed the finish line together, the whole Ford team would win. They could stage a tie and Miles and McLaren wouldn't have any reason to race each other to the end and risk crashing or blowing their engines. The #2 Ford was only a lap behind the #1 car.

"Oh hell," Shelby said, "let's do it that way then."

Beebe asked Bill Reiber, president of Ford-France, to talk to the officials and see if a tie could be arranged. There'd never been a dead heat at Le Mans before. Reiber did some investigating and returned.

"Leo," he said, "the officials say if you want to do it, they can arrange a tie and they will cooperate with you."

What did the boss think? Beebe had a chat with Henry II and Mr. Ford approved. What a statement they would make! In the pit, McLaren stood readying himself for his last shift. Shelby approached him with the idea of a dead heat, and the pilot was pleased. If it weren't for the trouble McLaren had with those Firestones at the start, and seconds lost with tire executives arguing in the pit over which rubber his car should be riding, he could've been in the lead anyway. Politics had slowed him down, and he had as much right to the win as anyone.

"Why *don't* you bring the cars over the line together?" he said in his New Zealand accent. "It would be much better public relations. Get a good picture."

At 2:46 P.M., Chris Amon slowed the #2 Ford into the pit for a final stop. His teammate McLaren was there, ready to jump in and finish the race. McLaren explained the situation.

"Who's supposed to win?" Amon asked.

"I don't know," McLaren answered, cryptically. "But I'm not going to lose."

At 3:01 P.M., Miles's teammate Denny Hulme pulled the leading car into the pit, where Miles was waiting, readying himself to bring the winning car home. The crew checked the tires, brakes, and oil carefully. While they added fuel—27 gallons, enough to get the car to the finish line—Miles and Hulme leaned back against the pit wall, both wearing their helmets. Shelby walked up behind them. He put his left hand on Miles's shoulder and spoke softly, ordering Miles to slow his pace and let McLaren catch up. Miles listened, staring through his sunglasses at the pavement beneath his feet. His body ached from exhaustion. The clouds had thickened and it appeared once again that the sky was ready to let loose on the crowd. When Shelby was done talking, Miles nodded his head.

"So ends my contribution to this bloody motor race," he shouted. He pulled off his shades and threw them across the pit. Heads turned. The mechanics knew something was up, but they didn't know what. All they knew was that Miles was suddenly furious. When Miles climbed into the cockpit, one of the crew leaned in.

"I don't know what they told you," he said, "but you won't be fired for winning Le Mans."

Miles threw the car in gear and rode off. Shelby watched him pull away. Miles was alone in the cockpit, with no one to control him. His destiny lay in his own hands and his right foot. He could keep up his speed and become champion of the greatest car race on earth, bring in the victory of a lifetime, a victory he had *earned*. He'd done the development work on this car. He'd won Daytona and Sebring. Victory would make for the triple crown. Or he could slacken and fall back into the dead heat. Give away the lead at Le Mans.

In the pit, the executives watched Miles, keeping their eyes on their stopwatches, waiting to see what he would do. Miles made those suits terribly nervous. On the track he kept an easy pace, and slowly the #2 Ford began to catch up. Miles lapped 10 seconds slower than McLaren, who was inching his way forward, gaining on the leader.

It would be a tie.

Suddenly Ford of France's Reiber appeared again before Beebe with an urgent message. "Leo, the officials now say a tie isn't possible." Reiber explained that, apparently, according to the rules, McLaren's car had qualified slower, so he had started farther back. Which meant at the finish, if there was a tie, McLaren's car would have traveled the farther distance over 24 hours. McLaren would win by about 20 feet.

"Oh my God," Beebe said, "that's not what we want at all. Is there any basis for appealing that?"

"No," Reiber said.

Beebe found himself in an excruciating predicament. What were they going to do now? They could call Miles and McLaren back in, inform them of the rules, and let them race it out. But that would be too risky. They could call them back in and tell McLaren to slack off so Miles would win. But Miles had earned a reputation. Hadn't he ignored pit signals to slow down, both here and at Sebring in March? Miles had rubbed some of the suits the wrong way. McLaren had been with the program since day one, long before Miles and Shelby, and he was one of the most well-liked men in racing. If Beebe let Miles win, he'd be giving the race to a guy that hadn't always been a team player. Miles was, however, an American citizen, which Beebe had to factor into the equation. There were innumerable angles to consider. All the while, the clock's minute hand rounded toward 4:00 P.M.

In the end, Beebe decided to do nothing. It was too late.

The excitement mounted during the final laps, as it always did at Le Mans. Spectators watched McLaren's black and silver machine stalk the leader until, as one in the crowd put it, "the car was cruising dead astern like a great black shark waiting for the kill." The announcer spouted wildly, unaware of the prefinish arrangements. Henry II watched and waited. The Dutray clock struck 4:00 P.M. A light rain began to fall, and in the pit the crews and executives stood on their toes in the mist so they could see the leaders when they rounded the White House bend in the final lap. The millions spent, the lives lost—all of it had led to this, America's finest hour in international racing.

They appeared—three cars riding in a tight pack, moving slowly toward the checkered flag. Miles and McLaren flanked each other in the #1 and #2 Shelby American Fords, and just behind Dick Hutcherson was piloting a Holman Moody Ford in third, #5 in gold with black trim. Hutcherson was twelve laps behind, almost exactly 100 miles. Neither Miles nor McLaren knew of the ruling, that McLaren would be declared the winner on a technicality. As they drove next to each other, moving perhaps at just 40 mph, both must have believed that they still had a chance at the lone victory. Hands on the wheel, feet on the pedal, all either had to do was step down for one moment, speed ahead, and the checkered flag would be their own for eternity. Perhaps they made eye contact in those final seconds. As a race official moved into the middle of the lane to wave the flag, McLaren suddenly moved forward, ahead of Miles. But it made no difference.

The checkered flag waved and the crowds swarmed onto the pavement. It was, to employ the great sports cliché, pandemonium. The drivers had to be careful not to run anyone down. Miles's teammate jumped into the car's passenger seat, keeping the door open and waving at the fans. They believed they had tied for the win and Miles moved the car slowly toward the podium. French officials stopped him, explaining that Miles had finished in second place. There was confusion. Miles's crew chief ran up to the car.

"I think I've been fucked," Miles told him.

For some minutes, the crowds grew more confused. What was happening? Who had won? In the broadcast booth, the ABC commentator scratched his head on camera. The officials didn't seem to know what was going on.

"Let's face it," Phil Hill said, "they know exactly what's going on down there. *We're* the ones who are confused."

Finally, it became clear to all who the official winners were. Henry II stepped onto the podium along with McLaren and Amon. The Kiwi drivers were handed huge bouquets of flowers and magnums of Moët. They stared awkwardly at the crowds. This was not how they intended to win. But in the official books, their victory was logged. They had traveled 3,009.6 miles at an average speed of 131 mph (including pit stops and all those slow corners), faster and further than anyone had ever gone before at Le Mans. Officials impounded their car and checked through a list of regulations, prodding its innards and taking fluid samples as if performing a mechanical autopsy.

As the ecstatic "Star-Spangled Banner" blared from the loudspeakers, the crowds cheered the winning drivers and waved flags of all kinds in the air. The Deuce raised a glass of champagne to his lips and in that stony face of his—which so rarely revealed any emotion in public—the world saw a genuine smile creep up and lock in.

Alone in the crowd, Miles stood in the drizzle, flabbergasted. He was wearing a Goodyear jacket now and a striped toque, and he held not a flute of champagne but a glass of Heineken. His face was crunched into a look of awe and gut-wrenching frustration. Cameras snapped at him and he could do nothing but hold his glass up, his tired eyes blank. He had indeed been fucked. He'd lost by some 20 feet. It was later that Miles saw McLaren face to face. The two men stared at each other. Miles grabbed the new Le Mans champion in his arms and gave him a bear hug. The race was over.

McLaren and his teammate Amon were given a little taste of what it was like to be in the Deuce's good graces. One VIP party followed another. Their pictures appeared in newspapers all over the globe, standing on the winners' podium with Henry Ford II.

"From there on," McLaren later recalled, "it was checkered flags, Champagne, and chauffeur-driven cars. Chris and I were flown to the States to attend Ford receptions. The enormous Lincoln that collected us at New York airport was flying a New Zealand flag on one wing and the Stars and Stripes on the other. In the plush drawing-room about 30 feet behind the engine, we swept through New York, feeling very much like the reigning monarch with all the peasantry peering in. Now I *know* how tough it must be at the very top."

Miles returned to his little cottage on Sunday Trail in the Hollywood Hills, with its rugged floral gardens and the cat that knew how to sit on the toilet. Everyone knew he was embittered. "He just couldn't get over it," his teammate Denny Hulme later said. Miles declined to give many interviews, but he did talk to Bob Thomas of the *Los Angeles Times*, who he knew well. "I considered we had won," Miles said about the now infamous photo finish. "But we were placed second by a technicality. I feel the responsibility for this rests with the decision by Ford, over my protests, to make the finish a dead heat. I told them I didn't think it would work." After the interview, Miles pleaded with the reporter. "Robert," he said, "please be careful how you report what I have said. I work for these people. They have been awfully good to me."

There wasn't much more to do or say. One summer morning, Miles awoke, dressed himself, got in his car, and went back to work.

Appropriately, Henry II had the last word on Le Mans in 1966. The world was changing. Communist countries aside, borders no longer existed when it came to industry. This was a new era unfolding. "We don't want to buy Ferrari anymore," Henry II told one reporter before leaving France. "Now we fear most of all the Japanese."

THE END OF THE ROAD

JUNE–AUGUST 1966

PARANOIA ATE AT John Surtees as he hurtled down the motorway in his Ferrari 330 GT, en route to meet with Mr. Ferrari. It was Wednesday, three days after the Le Mans finish. He had a press agent named Eoin Young with him. "I felt I ought not to be alone," Surtees later explained, describing his state. He had broken his contract with Ferrari, and it was imperative that he get the kind of press he wanted. *He wasn't fired; he quit.*

After they crossed the border into Italy, the pair pulled off the motorway into a roadside café to get some coffee. When Surtees walked into the café and got in line, the patrons spotted *Il Grande John*. As he stood there, more and more people recognized him. Then people rose to their feet and began to applaud for the British pilot. Had they not read their newspapers? The story was all over the place! But Surtees was a champion. He was used to the treatment. At this moment, however, the irony of it all stabbed at him.

Surtees and Young finally reached Maranello. Side by side they walked through the factory gate, their shadows trailing them on that bright summer day. Surtees was dressed as though he were entering a courtroom—dark business suit and tie, with a scuffed leather briefcase. They were met by Mr. Ferrari's secretary Valerio, then Gozzi appeared. And finally, Ferrari himself.

The driver and his boss closed themselves into Ferrari's office. These two men had always shared great respect. There would be no shouting. Surtees put forth his case: The team was throwing away opportunities because of absurd machinations and politics, he argued. He laid it all right there on Ferrari's desk, as perhaps no man ever had. What else was said in that office, only Surtees and Ferrari would ever fully know. "He made comments that explained a lot," Surtees later recalled, "things that he never talked about again during his lifetime, so I am certainly not going to raise them after his death."

The two agreed they would have to part ways, a "divorce" in Ferrari's words. They had shared so many victories, and all the racing pundits would forever agree that, politics aside, they would have dominated Formula One in the 1960s together. But it was not to be.

Ferrari and Surtees emerged from the office. Gozzi and Young joined them and they did what anyone would have done in their shoes: they went across the street to the Cavallino and had a drink.

Two months later, in the California desert, August 17, 1966. In the pit at Riverside, Ken Miles took refuge from the cruel sun in the shade of a canopy. Shirtless, he squinted and scratched at the graying hairs on his chest. He'd spent the morning tearing up Riverside's lap record in an experimental Ford racer called the J Car. In spite of Nader's safety crusade, Henry II had recently announced a $10 million racing budget for 1967.

They were going back to Le Mans. Miles would have another shot at the checkered flag.

He'd driven exceptionally fast that day. Though the cockpit baked in the August heat, he was feeling strong and energetic. All those morning jogs through the Hollywood Hills had paid off. As he rested in the shade, preparing for one last run of the day, the crew checked over the car and changed the wheels. In the parking lot, Miles's son Peter and a friend were messing around in a rental car. Behind the pits, a construction project was under way. With all the money flooding racing in 1966, Riverside was getting a $9 million makeover, with two new garages and offices to be occupied by Goodyear and Firestone.

When he was ready, Miles climbed back into the car, strapped his seatbelt, and took to the track. On the last lap of the last run of the day, he was burning down the mile-long backstretch at roughly 180 mph. Headed toward turn nine he slowed the car to about 100 mph. As he neared the bend, he was speeding no faster than one might on an empty highway, on a stretch of pavement as straight as a Roman road. A fireman was stationed at turn nine, watching Miles make his approach. In the pit, crewmembers were thumbing stopwatches when they heard the telltale screech of rubber.

"Oh my God," yelled one.

The J Car inexplicably veered sharply to its right and took flight, tumbling down a 10-foot embankment. The sound of bludgeoning metal ended with a final impact, after which the wreck exploded in flames.

Instantly, the pit emptied and by the time the crew reached the scene, firemen were already aiming extinguishers into the wreck. Miles lay 15 feet away, on his back. The crash was so violent, his seatbelt had torn from its mount and he'd been hurled to the ground. A quick glance was all it took to know that he'd been killed instantly.

Miles's son Peter arrived out of breath; he'd seen the fireball from the parking lot. "I remember seeing the car burning," Peter Miles later said. "But I didn't see my dad."

"Where is he?" Peter shouted.

A crewman pointed and Peter saw his father's body.

The trip home was a long one for Peter. He got stuck in traffic and he sat there, waiting for the congestion to loosen. When he arrived home at the little cottage on Sunday Trail, cars were already parked out front. The place was filled with friends and colleagues who'd heard the news.

In the morning the headlines appeared—Ken Miles was dead. Through the years, Miles had earned little except for a few dollars and some pats on the back from fans and colleagues. His acid tongue had made him some enemies, but nobody who knew Miles lacked respect for him. Only in his final year, at the improbable age of forty-seven, had he emerged as an international star. How gracefully he flew into turns, accelerating deeper into them than most others dared, the papers said. What skilled hands he had. He'd lived his life so close to the edge; that was his existence. Finally, he'd gone over that edge. As one obit put it, "We don't have to feel sorry for people who choose to live dangerously, and lose. So the bull wins one. The matador must take the risk. The closer he plays to the horn, the better the show . . . Well, Miles, good show."

Already the investigation was under way. When the J Car careened off the road, it behaved as if the wheels on one side simply locked up. It appeared that the rear of the car simply broke away. That someone would be blamed for Miles's demise was not something anyone wanted to face, but everyone knew instinctively that driver error was out of the question. Ford's aerospace division, Aeronutronic, was sending in a team to study the wreckage. Shelby was in Detroit at the time of the accident and he flew back immediately.

"We really don't know what caused it. The car just disintegrated," Shelby said, visibly shaken. "We have nobody to take his place. Nobody. He was our baseline, our guiding point. He was the backbone of our program." Then: "There will never be another Ken Miles."

Every piece of wreckage was examined as if it were a flight crash investigation. But the car was so severely destroyed by impact and fire, nothing could be proven, and Ford Motor Company had its image to uphold. "The evidence is that it was not mechanical," Don Frey said from his office in Dearborn. "We can't pinpoint anything which failed in the car before the crash."

In the end, all the evidence proved inconclusive. To this day, the cause of the accident that killed Ken Miles has never been determined.

On Saturday, August 20, at 2:30 P.M., a crowd of more than four hundred people filed into the Utter-McKinley Wilshire Mortuary at 444 South Vermont Avenue in Los Angeles. When everyone was seated, a family friend of the Mileses named Arthur Evans Sr. stepped up to the podium. Evans began a eulogy.

> We are here today to honor Kenneth Miles and to reassure his wife Mollie, and his son, Peter, of our continued devotion. Mollie has asked me to express her deep gratitude for your many kindnesses . . .

Shelby sat slumped in the church. All around him were friends and colleagues, a who's who of California sports car racing. Shelby had known Miles for a long time, and over the past four years they had formed a deep bond. As he listened to the eulogy, something burned inside him. It wasn't his bad heart; he didn't need a nitroglycerin pill. It was something else. His friend Miles loved machines and racing. He loved the instrumentation, the development work, the problem solving. And he loved winning.

> . . . Life, brilliant and vital as it may be, is an uncertain thing for everyone. It is in the memories of a life that we can find the immortal . . .

Ken Miles had won more than his share of trophies, Shelby was sure of that. In 1966 alone, he had won Daytona, and he had won Sebring. Le Mans would have meant the triple crown (a feat never achieved to this day). As the eulogy ended, and the crowd shuffled out the door into the bright California afternoon, Shelby kept coming back to the same thought: Miles should have taken that Le Mans trophy—the greatest of them all—with him to his grave.

EPILOGUE

> And I remember my grandfather saying, "Boy, we Americans, we can do anything."
>
> —Barack Obama at the Democratic National Convention, August 28, 2008

"Dear friends," Enzo Ferrari began his introductory note in the 1966 Ferrari yearbook. "This year we were finally beaten at Le Mans."

The grudge match between Ferrari and Henry Ford II continued in 1967. In the season's first face-off at Daytona in February, three Ferraris crossed the finish line together in one of the most dramatic victories in racing history. So spectacular was the finish, a photo of the three red cars speeding past the checkered flag at Daytona International Speedway today graces the wall in Ferrari's classic car restoration shop at the factory, blown up to nearly life-size.

But Enzo Ferrari's sports cars were unable to continue their dominance. The cars did not win at Le Mans in June. A Ferrari never won the 24 Hours of Le Mans again.

In 1969, Ferrari finally sold 50 percent of his company to Fiat. Ferrari's son Piero Lardi grew with the company. In 1978, he took on his father's last name and became a vice president, a title he still enjoys today.

Enzo Ferrari died on August 14, 1988, at the age of ninety. Upon his death, Fiat's stake in the company rose to 90 percent. The other 10 percent is still held by Piero Lardi Ferrari. Today, the Ferrari continues to be the most desired supercar in the world. The Scuderia Ferrari won its sixteenth Formula One constructor's World Championship in 2008, far more than any other team. In the rural fields where *Il Commendatore* built his factory during World War II, a bustling town has risen up around a modernized car plant. Across from the factory gate, the Cavallino still serves Ferrari's favorite dishes. Thousands of tourists visit Maranello annually, where the spirit of the man looms in every bottle of Lambrusco and every automobile that passes down the Via Abetone.

On June 12, 1967, Henry II claimed his second consecutive Le Mans victory. He climbed onto the podium along with his wife and the two winning drivers, Dan Gurney and A. J. Foyt. Gurney held a magnum of champagne. He aimed the bottle and—pop! "I drenched Henry Ford and his wife," he recalls, laughing. Thus, the victory tradition of spraying champagne after a race—de rigueur in Europe today—began. The race marked the first time an American car won Le Mans with a team of American drivers at the wheel. The *New York Times* called it an "All-American Victory" and "a cakewalk."

Two days later, Henry II launched a new company under the umbrella of his corporation: Ford of Europe, Inc., the first-ever pan-European automobile company. In the following years, when times grew tough for Ford Motor Company in America, Ford of Europe served as a huge profit center, the crutch that kept Henry II's empire moving forward.

Ford cars won Le Mans the following two years, making it four in a row (which did not top Ferrari's streak of six straight wins). In 1968 and 1969, the effort was led privately by J. W. Automotive Engineering, a team of Ford racing cars formed by John Wyer and sponsored by Gulf Oil.

The Deuce's quest to become the first American automobile manufacturer to conquer Le Mans was an experiment unlike any ever conducted in the world of modern automobiles, one that was life-defining for so many, and one that could never—under any circumstance—ever happen again. In the 1970s, the grandeur of the 24-hour classic began to fade. The race is still held every June, and still draws hundreds of thousands of spectators. But the glory days are past. Many fans believe the Ford-Ferrari rivalry represents automobile racing's true Golden Age.

Henry Ford II died on September 29, 1987, of pneumonia at Henry Ford Hospital. Even as he took his last breath, his name was on the building. The obituaries praised a larger-than-life maverick whose legacy lay as much in Europe as it did at home. The *Detroit News*: "He enjoyed beautiful women, strong drink." The *Financial Times*: "His insistence on setting up a highly integrated European operation had proved a brilliant success, so much so that the European plants far outshone their American sisters."

Born with hundreds of millions, Henry II could have lived a life of leisure and indulgence. Instead, he took a crumbling empire and led the effort to re-create it in a time of revolutionary change. At the time of his death, the corporation that bore his name was the most technically innovative and profitable automobile company in America.

Henry Ford II made cars.

In 1978, a bitter falling-out between Lee Iacocca and Henry II led to Iacocca's firing. He went on to take over Chrysler Corporation in 1979, which was on the brink of collapse. Ironically, Iacocca led a corporate comeback that eclipsed Henry II's achievements with Ford after World War II. It is, today, generally accepted as the greatest corporate comeback of all time.

John Surtees returned to Italy in 1967. *Il Grande John* won the Italian Grand Prix on Ferrari's home track not in an Italian car, but a Honda. "And the fans carried me high!" he remembers with a proud smile. Many watching at Monza understood the meaning behind this victory. The Japanese were about to release a full-fledged attack on the international automobile business.

Surtees left Honda to launch his own racing team, but his greatest successes were behind him. Today he lives quietly in England. He is still the only man in history to win Grand Prix World Championships on two wheels and four. Regarding the Ford-Ferrari business deal, he says, "I don't think Ferrari ever had any intention of going with Ford. I think it was a question of negotiating." And his controversial exit from Maranello? "Ferrari said to me before he died, 'John, we must remember the good times, not the bad,'" Surtees says. "I still love Italy and I love Italians and that's it."

Phil Hill retired from racing in 1967 and spent much of the next four decades restoring cars and writing. He died on August 28, 2008, at the age of eighty-one. Only one American after Hill has ever become Formula One World Champion: Mario Andretti, in 1978.

For the rest of the 1960s and into the 1970s, racing drivers continued to feed the fire with their lives. Lorenzo Bandini, Jimmy Clark, Ludovico Scarfiotti, Jo Schlesser, Bruce McLaren . . . At the end of the 1960s, ABC's *Wide World of Sports* aired a segment on the decade in racing. The names of the dead scrolled down the screen like so many teardrops as Jim McKay raised some difficult questions.

"We would not be reporting this decade in motor racing accurately if we did not mention the number of drivers who died on the racetracks of the world. The names you see [on screen] are only some of the more prominent men who lost their lives. You could make a case that motor racing is the most serious of all sports, or that it's not a sport at all. Like all sports, it carries a penalty for miscalculation. But here, the penalty is not a foul shot or 15 yards, but possibly your life. Whether this is sport, adventure, or foolhardiness depends on your personal definition."

In 1973, Scottish Grand Prix star Jackie Stewart retired at the height of his fame, and spearheaded a safety campaign. His efforts helped change emergency services and track facilities. When asked why he quit racing at the height of his fame, Stewart said, "I saw my friends killed and I saw what my wife was going through. I heard my son Paul ask when Daddy was going to die and I saw the nervous tic he developed, never knowing if I was going to come home."

Today, in part due to Stewart's efforts, death on the racetrack is a rare occurrence.

"You wouldn't believe what happens when a man like that turns on the faucet," Carroll Shelby says, looking back on the days of Henry II and Le Mans. It is springtime in 2007, and Shelby is sitting at his desk in his office in Gardena, California. The room is cluttered with toy cars. In the garage downstairs, someone's gunning an engine and the building's girders are soaking up the revs.

Against all odds, Shelby is alive. He is eighty-five years old. In the late 1960s, he got sick of the car business. He closed Shelby American and went to Africa. He made a comeback in the business in 1982, partnering with his old friend Iacocca to build high-performance Dodge cars. In 2004, Shelby teamed with Ford Motor Company again to produce a new line of Ford-Shelby Mustangs from his factory next to the Las Vegas Speedway. Today the man is made of used parts. He underwent a heart transplant in 1990; the organ came from a thirty-eight-year-old Vegas gambler. He has one of his son's kidneys.

After all these years, the controversial finish at Le Mans in 1966 still haunts Shelby. Former employees have even raised a conspiracy theory, that Ford executives had the French officials remove a lap from Ken Miles to put the tie into effect, that Miles had won that race all along. "I'll forever be sorry that I agreed with Leo Beebe and Henry Ford to have the three cars come across at the same time," Shelby says. "Ken was one and a half laps ahead and he'd have won the race. It broke his heart. Then we lost him in August."

The conversation moves along for another hour, and when our interview ends (there would be others), Shelby offers to drive me to the airport. We jump in a new Ford-Shelby Mustang GT-H. The car's got enough horsepower to outrun anything on the road, and though Shelby's eyes don't work so well, his right foot is still heavy. This is a man who won Le Mans wearing chicken-farmer overalls. As he weaves through the traffic on I-405, he says out of nowhere, "It broke my heart when we lost Ken." For a moment, there's silence. He sits in the driver's seat with his arms slightly bent, foot hard on the pedal. "I'm thinking about putting a million dollars in a deal to start a scholarship in his name," Shelby says. "I'm going to go down to the little town in East Texas where I was born. It has a wonderful college. I'm going to set up a Carroll Shelby school for mechanics. I'm going to put that together and have a scholarship in Ken Miles's name.

"I got to do something for him," Shelby says, squeezing the wheel. "I don't want him to be forgotten. I don't want him to be forgotten."

ACKNOWLEDGMENTS

This book could never have been written without the patience and time of all those who shared their thoughts and memories with me. With immeasurable reverence and humbleness, I want to thank (in no particular order) Carroll Shelby, Lee Iacocca, Edsel Ford II, Piero Ferrari, Franco Gozzi, John Surtees, Don Frey, Roy Lunn, Mario Andretti, Dan Gurney, Phil Hill, Peter Miles, Lloyd Ruby, A. J. Foyt, Robert Daley, Luigi Chinetti Jr., Richard Attwood, Phil Remington, Jim Hall, Jacque Passino, John Fitch, Homer Perry, Harry Calton, Bob Bondurant, and Eoin Young. Dave Friedman shared not just his stories, but also his tremendous photography.

Though this book has my name alone on the cover, it is the work of many Baimes. If ever there was one, this project was a family affair. My wife, Michelle, spent hours upon hours translating bad microfilm photocopies of Italian newspaper articles from the 1950s and 1960s. She was also the one that opened my eyes to the wonders of Italy. My mother, Denise, read every draft and spent many days in the New York Public Library hunting down obscure old articles and photographs. My father, David, also gave his thoughts on each draft and instilled in me as a youth the value of achieving goals, and also the deeper meaning of sport, which is what this book is all about.

Susan Canavan at Houghton Mifflin Harcourt is not only a great editor, but a terrific human being. Her thorough reads and help shaping this book (not to mention her encouraging spirit) deserve more than a little credit. Scott Waxman at the Waxman Literary Agency is the best agent in New York, a man any writer would feel blessed to have in his corner. Thank you also to Lucas Foster, Justin Manask, Byrd Leavell, Farley Chase, Alex Young, and Josh Bratman. The great car writer Ken Gross, who served as consultant on this project, is without a doubt the most respected journalist of any kind I have ever met. Thank you, Ken, for your close readings and for putting me in touch with some key sources. Your love of cars is an inspiration.

I would like to thank David "The Sage of the Stacks" Smith at the New York Public Library and Stephanie Epiro, a wonderful journalist in Italy who waded through old newspapers at the public library in Milan hunting down interviews with Enzo Ferrari. Special thanks go to Alan Hall at Ford Motor Company and Matteo Sardi at Ferrari, and, of course, to Chris Napolitano, Amy Grace Loyd, and all the brilliant editors at *Playboy*.

In terms of historical perspective, truth is in itself a kind of myth. Aldous Huxley once wrote, "The charm of history and its enigmatic lesson consist in the fact that, from age to age, nothing changes and yet everything is completely different." Through my many interviews, no two people ever told the same story. In many instances, the facts varied wildly. I put great effort into homing in on what really happened through the eyes of the men who lived through these exciting and tumultuous times. The dialogue was pulled directly from interviews and contemporaneous sources. Many of the men who appear in this book wrote memoirs themselves, which proved incredibly valuable. In some instances, I pulled characters' descriptions of events and used their own words as dialogue. For example, I have John Wyer saying at the Le Mans test weekend in 1964, "It's incredible that [Jo Schlesser] escaped with his life." Those are his exact words from his memoir, describing the moment he saw Schlesser's wrecked GT40.

The following books proved most valuable: *The Enzo Ferrari Memoirs: My Terrible Joys* by Enzo Ferrari; *The Fords: An American*

Epic by Peter Collier and David Horowitz; *Ford: The Dust and the Glory,* Volume 1, by Leo Levine; *John Surtees: World Champion* by John Surtees; *That Certain Sound* by John Wyer; *Memoirs of Enzo Ferrari's Lieutenant* by Franco Gozzi; *Ken Miles* by Art Evans; and *The Cobra Story* by Carroll Shelby. The collection of articles *Ferrari: The Man, the Machines,* edited by Stan Grayson, was extremely useful, as were the original Ford GT engineering papers from the mid-1960s, published today in one volume as *The Ford GT* by the Society of Automotive Engineers International. Dave Friedman's *Shelby GT40: The Shelby American Original Color Archives,* Brock Yates's *Enzo Ferrari: The Man, the Cars, the Races, the Machine,* and Robert Daley's *The Cruel Sport: Grand Prix Racing 1959–1967* (and his *New York Times* stories) were all extremely useful in culling facts and dialogue.

Mark Mandel at ABC was kind enough to retrieve the original *Wide World of Sports* broadcasts. My deep gratitude goes out to Jason Harper, Amy Grace Loyd, Robert Anasi, my wife Michelle for reading early chapters, and Bob Love, whose encouragement helped me to write the proposal for this book in the first place. Special thanks go to Scott Keogh and Jeff Kuhlman at Audi for their hospitality at Le Mans. I'd also like to thank my sister Abby, my Aunt Karen and Uncle Ken Segal, "the outlaws" Bill and Connie Burdick, Jack and Margo Ezell, Sydney Goldenberg, and Susan Baime for all their support through the years.

Finally, I'd like to thank a handful of people who taught me how to write: Jim Kaminsky, a great mentor who gave me all my initial breaks in this business, Chris Napolitano, Kenneth Silverman at New York University, Sarah Sherman and Brock Detier at the University of New Hampshire, my brilliant elementary school English teacher Mr. Breithaupt, and (though we've never met) Bruce Springsteen.

This book is dedicated to the lights of my life—Michelle, Clayton, and Audrey. I am the luckiest bastard on the planet.

NOTES

Introduction

x *A four hour sprint race*: Gordon H. Jennings, "Le Mans 24 Hours," *Car and Driver*, September 1964.

Prologue

xi *He shifted from third:* John Fitch (estimating speed and gearshifts at this point on the track), in discussion with the author.
He was wearing goggles: Mark Kahn, *Death Race: Le Mans 1955* (London: Barrie & Jenkins, 1976): 119–120.

xii *Each ticket had a warning:* Kahn, *Death Race*, 22.
Levegh had had a vision: Kahn, *Death Race*, 92–93.

xiii *It is too narrow:* Fitch, interview. Also: John Fitch with Art Evans and Don Klein, *Racing with Mercedes* (Redondo Beach, CA: Photo Data Research, 2005), 63.
Levegh was going about: Kahn, *Death Race*, 101.

xiv *He eyed a 16-foot-wide:* Kahn, *Death Race*, 104.

1. The Deuce

3 *I will build a motor car:* Henry Ford with Samuel Crowther, *My Life and Work* (Whitefish, MT: Kessinger, 2004), 73.
Just before 8:00 A.M.: William Serrin, "At Ford Everyone Knows Who Is the Boss," *New York Times Magazine*, October 19, 1969. In this piece, Henry II discusses his morning routine and commute.
"HFII"-monogrammed slippers: Peter Collier and David Horowitz, *The Fords: An American Epic* (San Francisco: Encounter Books, 2002), 260.
His commute took him west: Serrin, "At Ford."

4 *Ford Motor Company made:* Reference Document in the Ford Archives from "News Department," stamped June 8, 1961.

Henry II could see the stacks: Edsel Ford II, in discussion with the author.
Just before 10:30 A.M.: Booton Herndon, *Ford: The Unconventional Story of a Powerful Family, a Mighty Industry and the Extraordinary Men Behind It All* (New York: Avon Books, 1970), 208.
It was like being summoned: Lee Iacocca with William Novak, *Iacocca: An Autobiography* (New York: Bantam, 1984), 46.
We like what you are doing: Herndon, *Ford,* 208.
5 *One of the greatest marketing geniuses:* David Abodaher, *Iacocca* (New York: Zebra Books, 1984), 55.
If you want to be in this business: Herndon, *Ford,* 208.
The industry consumed 60 percent: Allan Nevins and Frank Ernest Hill, *Ford: Rebirth and Decline, 1933–1962* (New York: Charles Scribner's Sons, 1963), 436.
This is a nickel and dime: Robert Coughlan, "Co-Captains in Ford's Battle for Supremacy," *Life,* February 28, 1955.
My name is on the building: John S. Demott, "My Name Is on the Building," *Time,* October 12, 1987.
6 *Cost less per pound than a wheelbarrow:* "Auto, Upon Pound Basis, Ranks with Cheaper Machinery," *Washington Post,* October 17, 1926.
7 *He was a saint:* Collier and Horowitz, *The Fords,* 156.
This thing killed my father: Coughlan, "Co-Captains."
I'll take it only if: Collier and Horowitz, *The Fords,* 167.
8 *Can you believe it?:* George Koether, "How Henry Ford II Saved the Empire," *Look,* June 30, 1953.
Built 86,865 aircraft: Walter Hayes, *Henry: A Life of Henry Ford II* (New York: Grove Weidenfield, 1990), 24.
There were forty-eight plants: "The Rouge & the Black," *Time,* May 18, 1953.
The fat young man: Collier and Horowitz, *The Fords,* 163.
Beat Chevrolet: Coughlan, "Co-Captains."
Look, we're rebuilding: Collier and Horowitz, *The Fords,* 180.
9 *It was more than:* "Mrs. Ford Senior Sees Exhibit Here," *New York Times,* June 12, 1948.
Instrument of conquest: "Young Henry's $72,000,000 Gamble," *Newsweek,* June 14, 1948.
10 *The fight between Ford:* Coughlan, "Co-Captains."
Drive it home or make love to it: Clay Felker, "Iacocca: Whiz Kid, Senior Grade," *Esquire,* November 1962.
11 *Next year there will be:* Felker, "Iacocca: Whiz Kid."
The buyingest age group in history: "Ford's Young One," *Time,* April 17, 1964.
Iacocca was passed a photograph: Collier and Horowitz, *The Fords,* 249.
What we need is a campaign: "The Mustang—A New Breed Out of Detroit," *Newsweek,* April 20, 1964.
12 *The tomb:* "The Mustang."
Put in class for the mass: "The Mustang."
In European racing: "Press Box," *Detroit News,* February 27, 1961.
13 *The specific engine size:* "Ford Is Set to Ignore Industry Ban on Ads That Emphasize Speed," *Wall Street Journal,* June 12, 1962.
Chevrolet was funding a racing campaign: Alex Gabbard, *NASCAR's Wild Years:*

Stock-Car Technology in the 1960s (North Branch, MN: CarTech, 2005), 39. Also: Leo Levine, *Ford: The Dust and the Glory*, Volume 1 (Warrendale, PA: Society of Automotive Engineers, 2000), 265–267, 289–290.

14 *These guys are cheating:* Lee Iacocca, in discussion with the author.
Not since the heyday: "Thumping Year," *Newsweek*, April 2, 1962.
We believe this action: Levine, *Ford*, 267.
Never received a reply: Levine, *Ford*, 267.
GM's slice of the pie hit 61.6 percent: "G.M. Chunk of Auto Mart at 40 Year High," *Chicago Daily Tribune*, May 21, 1962.

15 *Now I don't want to imply:* Bob Ottum, "Someone Up There Said 'Let's Race,' and Lo, Ford Came Flying," *Sports Illustrated*, December 25, 1967.
But as others were doing it: Ford with Crowther, *My Life and Work*, 50.
Accordingly, we are withdrawing: "Ford Abandons Pact," *New York Times*, June 12, 1962.
Continue with unabated vigor: "Ford Junks Agreement to Ban Horsepower Ads," *Detroit News*, June 12, 1962.
We're going in with both feet: Levine, *Ford*, 290.
The biggest automotive scoop in years: Levine, *Ford*, 290.

16 *Detroit's romance with racing:* "Detroit Roaring Back Into Race Competition," *New York Times*, April 5, 1963.

2. Il Commendatore

17 *I am convinced:* Enzo Ferrari, *The Enzo Ferrari Memoirs: My Terrible Joys* (London: Hamish Hamilton London, 1963), 139.
Small chamber with a desk: Today, Ferrari's original *scuderia* office is on display at the Ferrari museum in Maranello.

18 *The puppy:* "Il Grande Anno del Costruttore dei Famosi Bolidi di Maranello," *Corriere della Sera*, February 28, 1958.
His diuresis, the presence of albumin: Ferrari, *Ferrari Memoirs*, 44.

19 *Don't let it get you down:* Ferrari, *Ferrari Memoirs*, 42.

20 *Ferrari's aim:* Winthrop Sargeant, "The Terrible Joys," *The New Yorker*, January 15, 1966.
The heartbeat of the creature: Gino Rancati, *Ferrari, the Man* (Milan: Sonzogno Publishing, 1977), 18.
Absurdly gifted artisans abound: Griffith Borgeson, "The Great Agitator," *Ferrari: The Man, the Machines*, ed. Stan Grayson (Princeton, NJ: Automobile Quarterly Publications, 1975), 13.

21 *It is my opinion:* Ferrari, *Ferrari Memoirs*, 26–27.
An agitator of men: Borgeson, "The Great Agitator," 23.
Sei di Denari: Franco Gozzi, *Memoirs of Enzo Ferrari's Lieutenant* (Milan: Giorgio Nada Editore, 2002), 10.

22 *Extremely emotionally attached:* Ferrari, *Ferrari Memoirs*, 27.
Speed-bewitched recluse: "Ferrari: Speed-Bewitched Recluse," *New York Times*, June 8, 1958.

23 *What life means:* Ferrari, *Ferrari Memoirs*, 44.

I have lost my son: Ferrari, *Ferrari Memoirs,* 44.
Racing no longer had meaning: Rancati, *Ferrari, the Man,* 100. Also, Brock Yates, *Enzo Ferrari: The Man, the Cars, the Races, the Machine* (New York: Doubleday, 1991), 242.
The office Dino kept: Hans Tanner and Doug Nye, *Ferrari* (Newbury Park, CA: Haynes Publishing, 1986), 7.
24 *Record del Miglio:* Gozzi, *Ferrari's Lieutenant,* 12–13. Also, Ferrari, *Ferrari Memoirs,* 18.
25 *Ferrari, why don't you put:* Ferrari, *Ferrari Memoirs,* 40.
50 percent car, 50 percent driver: Rancati, *Ferrari, the Man,* 27.
Near superhuman courage: Ferrari, *Ferrari Memoirs,* 61.
Almost a third of the scuderia*'s income:* Tanner and Nye, *Ferrari,* 22.
At the first bend: Ferrari, *Ferrari Memoirs,* 60.
26 *Huge sums of Reichmarks:* Yates, *Enzo Ferrari,* 85.
27 *Brauchitsch has burst:* Yates, *Enzo Ferrari,* 104.
28 *The future is here:* Gozzi, *Ferrari's Lieutenant,* 45.
If Chinetti was willing: Ken Gross (automotive historian and long-time contributing editor at *Playboy* magazine; Gross knew Chinetti and had heard him tell this story), in discussion with the author.
West 49th Street: Albert R. Bochroch, "Ferrari in America," *Ferrari,* ed. Stan Grayson, 177.
All we wanted to do was: Ferrari, *Ferrari Memoirs,* 41.
29 *The first Ferrari arrived:* Stan Nowak, "The First Ferraris," *Ferrari,* ed. Grayson, 100.
Roman suits of armor: Stan Nowak, "Bodies Beautiful—Ferrari Coachbuilding," *Ferrari,* ed. Grayson, 137.
The car which I have: Tanner and Nye, *Ferrari,* 7.
On December 2, 1956: Rancati, *Ferrari,* 106.
30 *I like the feeling of fear:* "A Man Who Was Born 400 Years Too Late," *Life,* April 9, 1956.
Making love is: Ken Purdy, "Portago," *Sports Cars Illustrated,* August 1957.
Burst into song: Ferrari, *Ferrari Memoirs,* 43.

3. Total Performance

31 *You go to a big:* Bob Ottum, "Someone Up There Said 'Let's Race,' and Lo, Ford Came Flying," *Sports Illustrated,* December 25, 1967.
A bitter controversy: "Controversy Brews Over Racing Issue," *Los Angeles Times,* February 10, 1963, and "Is Today D-Day at Daytona Race?" *Los Angeles Times,* February 24, 1963.
32 *It was all a smokescreen:* Kenneth Rudeen, "Big Smokescreen in Daytona," *Sports Illustrated,* March 4, 1963.
33 *In the open test:* "Advertising: New Horsepower Race," *New York Times,* March 7, 1963.
Ford Motor Companies: Internal company document, Ford Archives at Benson Ford Research Center, Acc. 1608, Box 1.
34 *If racing sells cars:* "If Racing Sells Cars . . . ," *Newsweek,* June 10, 1963. Also: Lee Iacocca, in discussion with the author.

Son of a gun, Lee!: Ottum, "Ford Came Flying."
Forget all the details: Ottum, "Ford Came Flying."

35 *The idea is staring:* Carroll Shelby, filmed interview, *The Cobra-Ferrari Wars* (Spirit Level Films, 2004).
Give him the money: This is a widely repeated story. For the sake of a source: Pete Brock, "The Car that Lived Up to Its Legend," *Car and Driver*, July 2001.

36 *You know, I think:* Carroll Shelby as told to John Bentley, *The Cobra Story* (New York: Trident Press, 1965), 195.
There's no man born: Shelby, *The Cobra Story*, 8.
Drive it?: Shelby, *The Cobra Story*, 35.

37 *A rich patron named:* Shelby, in discussion with the author.
I'm going to buy: Shelby, *The Cobra Story*, 74.
You can drive some: Shelby, interview. Also, Shelby, *The Cobra Story*, 88.

38 *Like a knife being stuck:* Shelby, interview.

39 *It seems that in a ridiculously:* "Corvette vs. Cobra: The Battle for Supremacy," *Road & Track*, June 1963.
Hi, I'm Lee: Dave Friedman (Shelby American official photographer), in discussion with the author.
I'm not an engineer: Shelby, interview. Also: Steve Smith, "The Ford in Carroll Shelby's Future," *Car and Driver*, June 1963.

40 *Does winning automobile races:* "Racing Victories Spur Ford Sales," *New York Times*, September 7, 1963.
I came to Europe: "European Investment by Ford Nears $800 million," *Los Angeles Times*, June 17, 1963.

41 *170 gigantic presses:* Factory descriptions from David Burgess-Wise, *Ford at Dagenham: The Rise and Fall of Detroit in Europe* (Derby, U.K.: Breedon Books Publishing Ltd., 2001), 143–168.

43 *530 Park Avenue:* Victor Lasky, *Never Complain, Never Explain: The Story of Henry Ford II* (New York: Richard Marek Publishers, 1981), 101.
Looked better in a bikini: Lasky, *Never Complain*, 101.
I've got the company: Collier and Horowitz, *The Fords*, 245.
Don't give me this shit: Collier and Horowitz, *The Fords*, 241.
I'm leaving: Lee Iacocca with William Novak, *Iacocca: An Autobiography*, 68.

44 *He was like a time bomb:* Collier and Horowitz, *The Fords*, 223.

4. Ferrari, Dino, and Phil Hill

45 *I was just a young man:* Phil Hill, *Ferrari: A Champion's View* (Deerfield, IL: Dalton Watson Fine Books, 2004), 85.
Jinx: Robert Daley, "The Ferrari 'Jinx,'" *Esquire*, November 1959.
It is the race: Steve McNamara, "Enzo Ferrari," *Sports Car Illustrated*, September 1959.

46 *Enough with these absurd:* "Proposta la Soppressione della Corsa a la Revisione di Tutte le Gare Su Strada," *Corriere della Sera*, May 13, 1957.
Enzo Ferrari, born in Modena: Enzo Ferrari, *The Enzo Ferrari Memoirs: My Terrible Joys* (London: Hamish Hamilton London, 1963), 156.

Why should I continue: Gino Rancati, *Ferrari, the Man* (Milan: Sonzogno Publishing, 1977), 145.

47 *One estimate put the figure:* Ken W. Purdy, "Stirling Moss: A Nodding Acquaintance with Death," *Playboy*, September 1962.

The driver pulled from a wreck: Purdy, "Stirling Moss."

Quitting was in itself: Purdy, "Stirling Moss."

Only those who do not move: Robert Daley, *Cars at Speed* (New York: Collier, 1966), 33.

For an instant he: "Jean Behra Killed in Race Crack-Up," *New York Times*, August 2, 1959.

48 *The modern racing car:* Piero Taruffi, "Stop Us Before We Kill Again," *Saturday Evening Post*, November 16, 1957.

Almost a superhuman quality: "Il Grande Anno del Costruttore dei Famosi Bolidi di Maranello," *Corriere della Sera*, February 28, 1958.

A modernized Saturn: "Vatican Paper Asks End of Auto Racing Deaths," *New York Times*, July 10, 1958.

His inner circle saw him struggle: Franco Gozzi, in discussion with the author.

I wouldn't want to be: Daley, "Ferrari 'Jinx.'"

Everything that I've done: "Il Grande Anno."

49 *It does not seem:* Griffith Borgeson, "The Great Agitator," *Ferrari: The Man, the Machines*, ed. Stan Grayson (Princeton, NJ: Automobile Quarterly Publications, 1975), 39.

How would you like: Phil Hill, "A Championship Season and Other Memories," *Ferrari*, ed. Stan Grayson, 220.

51 *Third or fourth Ferrari ever:* Phil Hill, in discussion with the author.

52 *Guaranteed not to cause ulcers:* William Nolan, *Phil Hill: Yankee Champion* (Carpinteria, CA: Brown Fox, 1996), 61.

You go as fast: "Racers Challenge Death in Mexico," *Life*, December 7, 1953.

On the end of the bench: Nolan, *Phil Hill*, 76.

53 *One of consummate and meticulous:* Pat Jordan, "Of Memory, Death, and the Automobile," *The Best Sports Writing of Pat Jordan* (New York: Perseus Books, 2008), 189.

I would so love to get: Robert Daley, *The Cruel Sport: Grand Prix Racing 1959–1967* (St. Paul, MN: Motorbooks International, 2005), 35.

He didn't have much contact: Hill, "Championship Season," *Ferrari*, ed. Grayson, 225.

Black-framed portraits: Robert Daley, in discussion with the author (Daley interviewed Ferrari in his office in 1958 with Hill present).

54 *We can win this race:* Daley, *Cars at Speed*, 128.

5. The Palace Revolt

55 *This kind of love:* Brock Yates, *Enzo Ferrari: The Man, the Cars, the Races, the Machine* (New York: Doubleday, 1991), 63.

I want to create a car: "Too Slow, You Lose—Too Fast . . . ," *Newsweek*, July 17, 1961.

57 *Mickey Mantle in a Ferrari:* "The Law and Mr. Phil Hill," *Los Angeles Times,* March 27, 1961.
He is resolve: Diana Bartley, "High Speed, High Brow," *Esquire,* June 1961.
Of course I am: "Too Slow . . . ."

58 *Maybe you ought:* "Enzo Ferrari, Builder of Racing Cars, Is Dead at 90," *New York Times,* August 16, 1988.
In front of his television: The author can only assume in this incredibly important race that Ferrari watched.

59 *And Trips?:* "Il Trionfatore P. Hill s'era Immaginato Tutto," *La Gazzetto della Sport,* September 11, 1961.
They were watching Von Trips's crash: Robert Daley, interview.
Are you going to quit: Robert Daley, "Why Men Race with Death," *New York Times Magazine,* October 1, 1961. Also: Robert Daley, interview.

60 *Fifteen Dead:* "Quindici I Morti per la Sciagura a Monza Cominiciata l'Inchiesta all'Audodromo," *Corriere della Sera,* September 12, 1961.
The man who could tell: "Race Driver Accused," *New York Times,* January 2, 1962.
You can't imagine: Nigel Roebuck, "Legends," *Motor Sport,* December 1997.

61 *We got rid of:* Yates, *Enzo Ferrari,* 295.
An assassin: Yates, *Enzo Ferrari,* 261.

6. Ferrari/Ford and Ford/Ferrari

62 *The American really:* William Faulkner, *Intruder in the Dust* (New York: Vintage International, 1991), 233.
Small, but nevertheless: Leo Levine, *Ford: The Dust and the Glory,* Volume 1 (Warrendale, PA: Society of Automotive Engineers, 2000), 499.
For what it is worth: Levine, *Ford,* 499.

63 *Filmer Paradise here:* Franco Gozzi, *Memoirs of Enzo Ferrari's Lieutenant* (Milan: Giorgio Nada Editore, 2002), 76.

64 *I never felt myself:* John Clinard, "The Day Ford-Ferrari Became Ford Versus Ferrari," *Car and Driver,* June 1974.
But be quite clear: Gozzi, *Ferrari's Lieutenant,* 78.
It would be like an: Roy Lunn, in discussion with the author.
The Drake was big: Gozzi, *Ferrari's Lieutenant,* 10.

65 *Ferrari led the Ford men into the works:* Factory description from Lunn, interview; Bob Bondurant (racing driver), in discussion with the author; and Brock Yates, *Enzo Ferrari: The Man, the Cars, the Races, the Machine* (New York: Doubleday, 1991), 301.
Owning one is: "Ferrari Superamerica 400," *Car and Driver,* April 1963.
Maranello's wind tunnel: Gozzi, *Ferrari's Lieutenant,* 78–79.
Boy is it clean: Lunn, interview.

66 *Frey and Ferrari spent hours:* Don Frey, in discussion with the author.
He drove like a mad man: Frey, interview.

67 *But here:* Gozzi, *Ferrari's Lieutenant,* 79.

But Mr. Ferrari: Clinard, "The Day."
My rights, my integrity: Clinard, "The Day."
A tirade that I: Gozzi, *Ferrari's Lieutenant,* 79.
68 *Let's go and eat:* Gozzi, *Ferrari's Lieutenant,* 79.
Mr. Ford, I failed: Frey, interview.
The longest lunch: Bob Ottum, "Someone Up There Said 'Let's Race,' and Lo, Ford Came Flying," *Sports Illustrated,* December 25, 1967.
We'll beat his ass: Frey, interview.
Prepare a presentation: Levine, *Ford,* 505.

7. Means and Motive

71 *To take control:* David Emory Shi, *The Bell Tower and Beyond: Reflections on Learning and Living* (Columbia: University of South Carolina Press), 97.
72 *Fiat had provided Ferrari:* "Il Grande Anno del Costruttore dei Famosi Bolidi di Maranello," *Corriere della Sera,* February 28, 1958.
73 *The Race of Truth:* Various authors, *Ferrari 1947–1997* (Milan: Giorgio Nada Editore srl, 1997), 118.
74 *The objective:* Roy Lunn, in correspondence with the author.
With the exception: Roy Lunn, "The Ford GT Sports Car," Society of Automotive Engineers (SAE) paper #670065.
75 *Attempting to meet:* Lunn, "The Ford GT Sports Car."
Amused at how quickly: Lunn, correspondence.
76 *Shelby loved to look:* Dave Friedman, in discussion with the author.
77 *That son of a bitch:* Michael L. Shoen, *The Cobra-Ferrari Wars 1963–1965* (Paradise Valley, AZ: CFW, 1988), 9. Carroll Shelby told the author this book was the most accurate of its kind.
There was something: Joan Sherman, on-camera interview, *The Cobra-Ferrari Wars* (Spirit Level Film, 2004).
Next year: John Jerome, "Carroll Shelby's Cobra Works," *Car and Driver,* September 1964.

8. Il Grande John

78 *The highlight of my career:* "Speed King Who Ruled Golden Age," *Daily Telegraph* (London), March 5, 2003.
What's so surprising: "Too Slow, You Lose—Too Fast . . . ," *Newsweek,* July 17, 1961.
79 *Just to sit in one:* Robert Daley, "That Blood-Red Ferrari Mystique," *New York Times Magazine,* July 25, 1965.
There is no finer: Daley, "Blood-Red Ferrari Mystique."
A noisy nirvana of automobiles: Brock Yates, *Enzo Ferrari: The Man, the Cars, the Races, the Machine* (New York: Doubleday, 1991), 281.
80 *Si, si . . . si, si:* Phil Hill, "A Championship Season and Other Memories," *Fer-*

rari: The Man, the Machines, ed. Stan Grayson (Princeton, NJ: Automobile Quarterly Publications, 1975), 234.

I wasn't sorry: William Nolan, *Phil Hill: Yankee Champion* (Carpinteria, CA: Brown Fox, 1996), 210.

It was a curious: John Surtees, ed. Alan Henry, *John Surtees: World Champion* (Surrey, U.K.: Hazleton Publishing, 1991), 107.

I would like you: Surtees, *John Surtees,* 107.

81 *Chief mechanic:* Surtees, *John Surtees,* 21.

82 *He doesn't seem:* Dennis May, "Changing a One-Track Mind," *Sports Car Illustrated,* September 1960.

Boy, if Horatio Alger: "World Champion: John Surtees," *Car and Driver,* February 1964.

A night at the hotel: Surtees, *John Surtees,* 136.

83 *We cannot compete:* John Surtees, in discussion with the author.

We are in a desert here: Surtees, interview.

84 *In the first act:* Enzo Ferrari, *The Enzo Ferrari Memoirs: My Terrible Joys* (London: Hamish Hamilton London, 1963), 48.

85 *The threat of Ford Motor Company:* Franco Gozzi, in discussion with the author; Surtees, interview.

9. The Ford GT40

86 *This is a year:* Henry Ford II, "Technological Exploration: Our Danger Is Overconfidence," *Vital Speeches of the Day* (Vol. 30, Issue 8), January 15, 1964, 246.

87 *Pappy can tell you:* John Wyer, *That Certain Sound* (Somerset, U.K.: Foulis Motoring Books, 1981), 117.

Don Frey was receiving letters: Roy Lunn, in discussion with the author.

None of which were available in America: A great majority of the reporting on car design came from *The Ford GT: New Vehicle Engineering and Technical History of the GT-40* (Warrendale, PA: Society of Automotive Engineers International, 2004). Other sources include Roy Lunn (interview and personal correspondence); Karl E. Ludvigsen, *The Inside Story of the Fastest Fords* (Turin, Italy: Style Auto Editrice, 1971); and Leo Levine, *Ford: The Dust and the Glory,* Volume 1 (Warrendale, PA: Society of Automotive Engineers International, 2000).

89 *Coordinates showed a top speed:* Roy Lunn, "The Ford GT Sports Car," SAE paper #670065.

90 *Well behind schedule:* Wyer, *That Certain Sound,* 127.

91 *Eleven months after:* Lunn, "The Ford GT Sports Car."

A racing car chassis: Bruce McLaren, filmed interview, *Birth of the Ford GT* (Motorfilms Quarterly, Volume 1, 2002).

Roy?: Lunn, interview.

92 *The car was flown:* Wyer, *That Certain Sound,* 130.

Feared the worst: Lunn, interview.

93 *At medium speeds:* Wyer, filmed interview, *Birth of the Ford GT.*
The press had called him: Philip Payne, "The Last Race of Count Crash," *Sports Illustrated,* September 18, 1961.

10. Loss of Innocence

94 *In the long run:* Eddie Merchant, "Racing's Haunted Driver," *Saturday Evening Post,* May 26, 1962.
An aggregate of only: Karl E. Ludvigsen, *The Inside Story of the Fastest Fords* (Turin, Italy: Style Auto Editrice, 1971), 14.
Geared to run at 200 mph: "Le Mans Trials Prove Ford Must Work to Catch Ferrari," *New York Times,* April 21, 1964.
95 *I can't believe this:* John Wyer, filmed interview, *Carroll Shelby: The Man and His Car* (Duke DVD, 1990).
96 *The car was fishtailing:* Wyer, filmed interview, *Carroll Shelby.*
It is incredible: John Wyer, *That Certain Sound* (Somerset, U.K.: Foulis Motoring Book, 1981), 134.
We wrecked both of them: Bob Ottum, "Someone Up There Said 'Let's Race,' and Lo, Ford Came Flying," *Sports Illustrated,* December 25, 1967.
97 *In trials that ended:* "Le Mans Trials Prove Ford Must Work to Catch Ferrari," *New York Times,* April 21, 1964.
We appreciate your coming: "Remarks by L.A. Iacocca; Mustang National News Conference, New York, New York, April 13, 1964," stamped Editorial Services Dept, Public Relations Staff and provided by the Benson Ford Research Center.
98 *A new breed out of Detroit:* "The Mustang—A New Breed Out of Detroit," *Newsweek,* April 20, 1964.
With its long hood: "Ford's Young One," *Time,* April 17, 1964.
Iacocca had dinner with: Lee Iacocca, in discussion with the author. Also, "Death Takes No Holiday," *Newsweek,* June 8, 1964.
99 *Jesus Christ!:* David Davis, "Indianapolis 500," *Car and Driver,* August 1964.

11. Le Mans, 1964

100 *I am hypnotized:* Franco Gozzi, *Memoirs of Enzo Ferrari's Lieutenant* (Milan: Giorgio Nada Editore, 2002), 112.
101 *462 pounds sterling:* John Wyer, *That Certain Sound* (Somerset, U.K.: Foulis Motoring Book, 1981), 117.
Had the effect: Roy Lunn, "The Ford GT Sports Car," SAE paper #670065.
Mr. Wyer doesn't have: Harry Calton (head of corporate communications for Ford's Le Mans effort), in discussion with the author.
Wyer posted timetables: Calton, interview.
102 *We are all shocked:* "Ford to Race in Spite of '500' Deaths," *Detroit News,* June 1, 1964.
Outside of the United States: Paul Evan Ress, "For Ferrari, Some Fast Company at Last," *Sports Illustrated,* June 29, 1964.

103 *To a firm like Ferrari:* John Surtees, interview by Stirling Moss, ABC *Wide World of Sports,* June 28, 1964.

104 *Leer at ABC's script girl:* Gordon H. Jennings, "Le Mans 24 Hours," *Car and Driver,* September 1964.
After all: Bob Ottum, "A New Race Produces an Old Ferrari Story," *Sports Illustrated,* February 24, 1964.
There were no dramatic meetings: John Surtees, in discussion with the author.
He wanted everything done precisely: Phil Hill, in discussion with the author.
We want to finish the race: Ress, "For Ferrari, Some Fast Company."

105 *June 11, 1955:* Robert Daley, *Cars at Speed* (New York: Collier Books, 1962), 123.

106 *His Cobra had clocked 197 mph:* Michael L. Shoen, *The Cobra-Ferrari Wars 1963–1965* (Paradise Valley, AZ: CFW, 1988), 169.

107 *Billiard table smooth:* John Surtees, edited by Alan Henry, *John Surtees: World Champion* (Surrey, U.K.: Hazleton Publishing, 1991), 131.
When you start [racing]: Surtees, *John Surtees,* 177.

108 *Word from the course:* ABC *Wide World of Sports* footage, June 28, 1964.
What are you doing up there?: Eoin Young (journalist and public relations man with Bruce McLaren), in discussion with the author. Also: Eoin Young, *McLaren Memories* (Newbury Park, CA: Haynes Publishing, 2005), 167.

109 *He was cruising at 185 mph:* Speeds, rpm readings, and gear selections are taken from Figure 1 in "The Ford GT Sports Car" (SAE paper #670065), a map of the Le Mans circuit with all the data prepared by Roy Lunn with the help of Ford Motor Company team racing drivers.

110 *True concentration:* Phil Hill, "Martini & Rossi Present the Phil Hill Competition Driving Lesson," *Car and Driver,* November 1962.
140 degrees Fahrenheit: Mario Andretti, in discussion with the author.
Not until you find: Hill, "Martini & Rossi Present."

111 *Well, for God's sake:* Young, *McLaren Memories,* 167.
Wildly ecstatic: Young, *McLaren Memories,* 167.
He had hit 210 mph: Young, *McLaren Memories,* 167.
Two minutes seven seconds had passed: Leo Levine, *Ford: The Dust and the Glory,* Volume 1 (Warrendale, PA: Society of Automotive Engineers, 2000), 514.
A Ferrari was like insurance: Michael L. Shoen, *The Cobra-Ferrari Wars 1963–1965* (Paradise Valley, AZ: CFW, 1988), 167.

112 *I saw some flames:* Richard Attwood, interview.
This was the result: Wyer, *That Certain Sound,* 140.
It is enough: Wyer, *That Certain Sound,* 140.
Irony eating at him: Roy Lunn, in discussion with the author.

113 *Driving at night:* Young, *McLaren Memories,* 166.
The best 500 racing miles: Young, *McLaren Memories,* 166.
In a tent: "Guichet and Vaccarella Drive Ferrari to 5-Lap Victory at Le Mans," *New York Times,* June 22, 1964.

114 *It's the middle:* ABC *Wide World of Sports* footage, June 28, 1964.

115 *A wonderful thing happens:* Ken Purdy, "Masten Gregory Lives," *Esquire,* January 1969.
There's nothing like: Young, *McLaren Memories,* 188.

116 *Why didn't we find:* Phil Hill, "A Championship Season and Other Memories," *Ferrari: The Man, the Machines,* edited by Stan Grayson (Princeton, NJ: Automobile Quarterly Publications, 1975), 221.
Brakes okay?: ABC *Wide World of Sports,* June 28, 1964.
117 *Congratulations Bob:* ABC *Wide World of Sports,* June 28, 1964.
118 *Thank you for beating:* Shoen, *The Cobra-Ferrari Wars,* 178.
Fourth isn't bad: Ress, "For Ferrari, Some Fast Company."

12. Aftermath

119 *Like an evangelist missionary:* John Wyer, *That Certain Sound* (Somerset, U.K.: Foulis Motoring Book, 1981), 141.
If he told me: Booton Herndon, *Ford: The Unconventional Story of a Powerful Family, a Mighty Industry and the Extraordinary Men Behind It All* (New York: Avon Books, 1970), 39.
120 *I don't know anything:* Leo Levine, *Ford: The Dust and the Glory,* Volume 1 (Warrendale, PA: Society of Automotive Engineers, 2000), 520.
You could lay it: Levine, *Ford,* 517.
Wyer was dumbfounded: Wyer, *That Certain Sound,* 141.
121 *Any right thinking Italian:* Denis Jenkinson, "Il Grande John," *Ferrari: The Man, the Machines,* edited by Stan Grayson (Princeton, NJ: Automobile Quarterly Publications, 1975), 272.
122 *Surtees would later come:* John Surtees, in discussion with the author.
What does Ferrari have: Wyer, *That Certain Sound,* 144.
That Ford smell: C. Gayle Warnock, *The Edsel Affair* (Paradise Valley, AZ: Pro West, 1980), 22.
123 *A way to solve the problem:* Roy Lunn, in discussion with the author.
More power is always: Wyer, *That Certain Sound,* p. 144.
124 *To get to the Bahamas:* John Horsman, *Racing in the Rain: My Years with Brilliant Drivers, Legendary Sports Cars, and a Dedicated Team* (Phoenix, AZ: David Bull Publishing, 2006), 67.
I don't know anything: Levine, *Ford,* 520.
125 *You want 'em?:* Dave Friedman, *Remembering the Shelby Years—1962–1969* (Los Angeles: The Carroll Shelby Children's Foundation, 2006), 290.
126 *It's bloody awful:* Dave Friedman, *Shelby GT40* (St. Paul, MN: Motorbooks, 2006), 18.
127 *Shaking so many hands:* John Surtees, ed. Alan Henry, *John Surtees: World Champion* (Surrey, U.K.: Hazleton Publishing, 1991), 157.
Formula 1 is for: Julius Weitman, "Enzo Ferrari Off Guard," *Car and Driver,* July 1965.
Closed, like a walnut: Griffith Borgeson, "The Great Agitator," *Ferrari: The Man, the Machines,* edited by Stan Grayson (Princeton, NJ: Automobile Quarterly Publications, 1975), 10.
After lunch, Surtees and Ferrari: Weitman, "Enzo Ferrari Off Guard."

128 *The facial expression:* Enzo Ferrari, *The Enzo Ferrari Memoirs: My Terrible Joys* (London: Hamish Hamilton London, 1963), 37.

13. Henry II, Shelby, and Daytona

131 *Grand Prix racing:* Robert Daley, *The Cruel Sport: Grand Prix Racing 1959–1967* (St. Paul, MN: Motorbooks International, 2005), 214.

132 *Fifty million went:* "'Golden Era' for Racing Just Around Next Turn," *New York Times,* April 4, 1965.

The sudden outpouring: "Hollywood Agog Over Auto Racing," *New York Times,* November 19, 1965.

Never before has: "Auto Industry Using Advertising to Keep the Wheels Turning," *Los Angeles Times,* February 2, 1964.

The company is now enjoying: "Car Tax Cut Next Logical Step—Ford," *Chicago Tribune,* May 21, 1965.

133 *Can you do it?:* Carroll Shelby, in discussion with the author.

Frey never dreamed: "Personality: From Slide Rule to Marketing," *New York Times,* February 21, 1965.

Six potential car: "Personality."

Mr. and Mrs. Henry Ford II: Statement from Bodman, Longley, Bogle, Armstrong & Dahling dated August 3, 1963, Ford Archives.

134 *Henry, look at:* William Serrin, "At Ford Everyone Knows Who Is the Boss," *New York Times Magazine,* October 19, 1969.

He was sitting in the back: John Wyer, *That Certain Sound* (Somerset, U.K.: Foulis Motoring Book, 1981), 148.

135 *We are taking this:* Steve Smith, "The Ford in Carroll Shelby's Future," *Car and Driver,* June 1965. Also: "Shelby to Build New Ford Racers," *Los Angeles Times,* January 28, 1965.

Official move was March 1: Dave Friedman, *Cobra: The Shelby American Original Archives 1962–1965* (Los Angeles: The Carroll Shelby Children's Foundation, 1998), 143.

136 *Up and down the runways:* Shelby, interview.

Shelby paces about: Coles Phinizy, "Snakes, Butter Beans, and Mister Cobra," *Sports Illustrated,* May 17, 1965.

You have to go 90 mph: Phinizy, "Snakes, Butter Beans."

Hello, butter bean: Phinizy, "Snakes, Butter Beans."

How would you like: Phinizy, "Snakes, Butter Beans."

137 *It may sound odd:* Leo Levine, *Ford: The Dust and the Glory,* Volume 1 (Warrendale, PA: Society of Automotive Engineers, 2000), 520.

Everyone wore boots: Dave Friedman, in discussion with the author.

Technicians from Aeronutronic: Tony Hogg, "A Look at the Daytona Winner Ford GT-40," *Road & Track,* May 1965. Also: Friedman, interview.

138 *Members of the team popped pills:* Carroll Shelby, filmed interview, *The Cobra-Ferrari Wars* (Spirit Level Film, 2004).

We have several advantages: Smith, "The Ford in Shelby's Future."

139 *The same careful workmanship:* Art Evans, *Ken Miles* (Redondo Beach, CA: Photo Data Research, 2004), 73.
His unit was among the first: Mollie Miles, "Miles Away!" *Road & Track,* November 1954.
His son Peter remembered: Peter Miles, in discussion with the author.
140 *Filthy army jacket:* Homer Perry (Ford's Le Mans Committee organizer), in discussion with the author.
I'd rather die: Friedman, interview.
Happiness Is a Hot Rod: Michael L. Shoen, *The Cobra-Ferrari Wars 1963–1965* (Paradise Valley, AZ: CFW, 1988), 253.
141 *Shaking with fear so bad:* Shoen, *Cobra-Ferrari Wars,* 254.
Ruby headed over: Lloyd Ruby, in discussion with the author.
142 *This is a team effort:* Shoen, *Cobra-Ferrari Wars,* 255.
143 *I got drunker than shit:* Shelby, interview.
Ruby could crawl: Three anonymous interviews.

14. 220 mph

144 *Women are more:* Winthrop Sargeant, "The Terrible Joys," *The New Yorker,* January 15, 1966.
145 *La polemica di Surtees:* "La Polemica di Surtees Favoriva Bandini?" *Gazzetta della Sport,* June 9, 1965.
146 *Naturally, from a commercial:* John Surtees, filmed interview, *The European Grand Prix 1964* (Terrific Stuff Videos, n.d.).
Ferrari Cars Quit: "Ferrari Cars Quit World Title Meets," *New York Times,* April 2, 1965.
147 *Do you mind:* Carroll Shelby, in discussion with the author.
I been losing money: Shelby, interview.
Dick Hall, an oilman who: Carroll Shelby's official Web site, "Timeline," http://www.carrollshelby.com/history.html (accessed September 1, 2008).
I don't give a: Shelby, interview.
148 *Watch out for my balls!:* Charlie Agapio, filmed interview, *The Cobra-Ferrari Wars* (Spirit Level Film, 2004).
Hall Ushers in New Era: "Hall Ushers in New Era at Sebring," *Los Angeles Times,* March 29, 1965.
149 *His angina was killing him:* Shelby, interview. Also: Carroll Shelby, filmed interview, *The Cobra-Ferrari Wars* (Spirit Level Film, 2004).
He drank a lot of liquor: Shelby, interview.
I got something I'd like: Roy Lunn, in discussion with the author.
150 *He essentially stole:* Lunn, interview.
It doesn't look: Lunn, interview.
152 *Tips of his fingers:* Art Evans, *Ken Miles* (Redondo Beach, CA: Photo Data Research, 2004), 87.
I suppose you'd: Leo Levine, *Ford: The Dust and the Glory,* Volume 1 (Warrendale, PA: Society of Automotive Engineers, 2000), 529.

What does everybody: Levine, *Ford,* 529.

153 *The Deuce arrived:* Shelby, interview.

Can you imagine: "When It Comes to Cobras, Call Shelby—a Real Charmer," *Los Angeles Times,* June 13, 1965.

154 *He distanced himself:* Enzo Ferrari, "Ricordo di Bruno," Official Ferrari Yearbook, 1965.

Deserti is really fresh: "La Ferrari Recupera Baghetti e Lancia Biscaldi e Deserti," *Corriere della Sera,* April 7, 1965.

It represents the: Ferrari, "Ricordo di Bruno."

155 *When such passion connected:* Ferrari, "Ricordo di Bruno."

He's out!: "Fuori Pista Una Ferrari Morto Il Giovane Deserti," *Corriere della Sera,* May 26, 1965.

156 *Enzo Ferrari believes:* Robert Daley, "That Blood-Red Ferrari Mystique," *New York Times Magazine,* July 25, 1965.

With the great increase: Dave Friedman, *Shelby GT40* (St. Paul, MN: Motorbooks, 2006), 58.

The news of Ford's new: John Surtees, in discussion with the author.

The pistons are as big: Piero Ferrari, interview (translated by Ferrari press liaison Matteo Sardi).

The two posed for a ceremonial: Friedman, *Shelby GT40,* 57.

15. Le Mans, 1965

157 *Once, in my racing days:* Enzo Ferrari, *The Enzo Ferrari Memoirs: My Terrible Joys* (London: Hamish Hamilton London, 1963), 90.

The car that won: John Lovesy, "It Was Murder Italian Style," *Sports Illustrated,* June 28, 1965.

Mid-season rift: "U.S. Challenge to Ferrari Domination," *Times of London,* June 15, 1965.

158 *Surtees believed that the car:* John Surtees, edited by Alan Henry, *John Surtees: World Champion* (Surrey, U.K.: Hazleton Publishing, 1991), 172.

159 *There comes a time when:* Robert Daley, "Sundown of a Champion," *Saturday Evening Post,* May 8, 1965.

160 *Sundown of a Champion:* Daley, "Sundown."

It's absolutely frightening: Karl Ludvigsen, *The Inside Story of the Fastest Fords* (Turin, Italy: Style Auto Editrice, 1971), 40.

If we could get it more stable: Lovesy, "Murder Italian Style."

Roy Lunn had an idea: Dave Friedman, *Shelby GT40* (St Paul, MN: Motorbooks, 2006), 63.

Let it out: Leo Levine, *Ford: The Dust and the Glory,* Volume 1 (Warrendale, PA: Society of Automotive Engineers, 2000), 533.

161 *Phil Hill of Santa Monica:* "Phil Hill Breaks Le Mans Record," *New York Times,* June 19, 1965.

Why did we pick: ABC *Wide World of Sports* footage, June 19, 1965.

Shelby was taking: Lovesy, "Murder Italian Style."

162 *I didn't think:* Ken Purdy, "Masten Gregory Lives!" *Esquire,* January 1969.

163 *Drivers take your positions:* Footage of the Le Mans 24 Hours, http://www.youtube.com/watch?v=tpkvvsnv7Xk, accessed November 25, 2008.

164 *This is sport that makes:* Mark Twain, *Life on the Mississippi* (Whitefish, MT: Kessinger Publishers, 2004), 173.
What do you think, Phil?: ABC *Wide World of Sports* footage, June 19, 1965.
6,000 rpm and 6,500 on straights: Carroll Smith, "Race Report Le Mans 1965," stamped "Special Vehicles Sports Car Manager Ford Division," unpublished official Ford Motor Company document, 5. Also: Eoin Young, *McLaren Memories* (Newbury Park, CA: Haynes Publishing, 2005), 195.
It's like a rocket ship!: Dave Friedman, in discussion with the author.

165 *I can smell the chicken:* Michael L. Shoen, *The Cobra-Ferrari Wars 1963–1965* (Paradise Valley, AZ: CFW, 1988), 339.

166 *You should have seen:* Friedman, *Shelby GT40*, 67.

16. Le Mans, 1965: The Finish and the Fallout

168 *Racing with Chinetti:* Michael L. Shoen, *The Cobra-Ferrari Wars 1963–1965* (Paradise Valley, AZ: CFW, 1988), 167.

169 *An elaborate, protracted:* Brock Yates, *Enzo Ferrari: The Man, the Cars, the Races, the Machine* (New York: Doubleday, 1991), 2.
Having his best year ever: Albert R. Bochroch, "Ferrari in America," *Ferrari, the Man, the Machines,* edited by Stan Grayson (Princeton, NJ: Automobile Quarterly Publications, 1975), 177.
The Old Man cheated Luigi: Yates, *Enzo Ferrari*, 239.

170 *Chinetti found himself, with flashlight:* Filmed interview with Carroll Shelby, Brock Yates, and Ken Gross, February 11, 2005, Saratoga Automobile Museum Living Legends Series (Saratoga Automobile Museum, 2005).
Ferrari factory representative appeared: Luigi Chinetti Jr., in discussion with the author.
You're going to tell: Chinetti Jr., interview.

171 *These are literally:* ABC *Wide World of Sports* footage, June 20, 1965.
It may seem odd to Americans: ABC *Wide World of Sports* footage, June 20, 1965.

172 *What the hell happened?:* Carroll Shelby, in discussion with the author. Also: Edsel Ford II, in discussion with author.
That's a good car!: Shelby, interview.
Headed to the nearest bar: Shelby, interview.
The greatest defeat ever: Karl Ludvigsen, *The Inside Story of the Fastest Fords* (Turin, Italy: Style Auto Editrice, 1971), 34.
Murder Italian Style: John Lovesy, "It Was Murder Italian Style," *Sports Illustrated,* June 28, 1965.
He could wind up: David E. Davis Jr., Chris McCall, and Al Bochroch, "24 Heures du Mans," *Car and Driver,* September 1965.
$6 million: John Wyer, *That Certain Sound* (Somerset, U.K.: Foulis Motoring Book, 1981), 150.

173 *I was very disappointed:* Wyer, *That Certain Sound*, 151.

I felt like I was: Booton Herndon, *Ford: The Unconventional Story of a Powerful Family, a Mighty Industry and the Extraordinary Men Behind It All* (New York: Avon Books, 1970), 24.
You got your asses: Yates, *Enzo Ferrari*, 318.
Ford wins Le Mans in 1966: Shelby, interview.
I wonder what our: Larry Edsall, *Ford GT: The Legend Comes to Life* (St. Paul, MN: Motorbooks, 2004), 8.
The shit hit the fan: Dave Friedman, in discussion with the author.
174 *Almost anticlimactic:* Carroll Shelby, filmed interview, *The Cobra-Ferrari Wars* (Spirit Level Film, 2004).
Before the summer was out: Original blueprints filed at the Ferrari factory classic car shop in Maranello, Italy, shared with the author in 2006.
175 *Red driving gloves:* Chinetti Jr., interview.
How many kids: Mario Andretti, in discussion with the author.
176 *In the classes we:* John Surtees, in discussion with the author. Also: John Surtees, *John Surtees: World Champion* (Surrey, U.K.: Hazleton Publishing, 1991), 158.
It's not right: Surtees, interview.
That's the last thing: Surtees, interview. Also: Surtees, *John Surtees*, 176.

17. Survival

179 *The Ferrari stands:* Barbara La Fontaine, "Miles: 'Fast Enough to Win, Slow Enough to Finish,'" *Sports Illustrated*, February 14, 1966.
He was bleeding: "Surtees Suffers Fractures, Back Injury in Crash," *Washington Post*, September 25, 1965.
180 *We don't want to:* John Surtees, in discussion with the author.
I'm sorry, John: Surtees, interview.
181 *You just get my car:* John Surtees, *John Surtees: World Champion* (Surrey, U.K.: Hazleton Publishing, 1991), 178.
I'll have a go: Surtees, interview.
I would think he will: "Driver Surtees Expected to Make Full Recovery," *Washington Post*, September 26, 1965.
182 *We are not really able:* Surtees, interview.
Don't let those: Surtees, *John Surtees*, 178.
An involuntary, and unwanted: Surtees, *John Surtees*, 179.
Time's getting on: Surtees, *John Surtees*, 179.
183 *He's big, isn't he?:* Surtees, interview.
We reckon you'll: Surtees, *John Surtees*, 180.
I felt as though: Surtees, *John Surtees*, 180.
We're going to try: Surtees, *John Surtees*, 180.
184 *Feeling very poor:* Surtees, *John Surtees*, 180.
In December, the hospital's hallways: Associated Press photograph, December 23, 1965.

185 *I have real confidence:* Pete Coltrin, "Ferrari 330P3 and Dino 206/S," *Road & Track,* April 1966.
Four liters are enough: Coltrin, "Ferrari 330P3 and Dino 206/S."

18. Rebirth

187 *You'd better win:* Homer Perry (Ford's Le Mans Committee organizer), in discussion with the author.
188 *Anything you want:* Leo Levine, *Ford: The Dust and the Glory,* Volume 1 (Warrendale, PA: Society of Automotive Engineers, 2000), 538.
189 *Miserable, slippery:* "Runaway at Daytona," *Time,* February 18, 1966.
Firemen and medical staff: Perry, interview.
I remember some really: Art Evans, *Ken Miles* (Redondo Beach, CA: Photo Data Research, 2004), 88.
190 *Socks, his old army jacket:* Perry, interview (the two occasionally roomed together on the road).
An almost mystical sense: Evans, *Ken Miles,* 80.
Brake fluid boiling: Shelby American Inc.—Engineering & Development Report, January 25, 1966, 3a.
The car was uncontrollable: Shelby American Inc.—Engineering & Development Report, January 25, 1966, 3aa.
Shelby and Miles flew out: Official Shelby American memorandum.
192 *Holman Moody this:* Carroll Shelby, in discussion with the author.
Someday you're going to: Dave Friedman, *Remembering the Shelby Years—1962–1969* (Los Angeles: Carroll Shelby Children's Foundation, 1998), 291.
The driver list reads: "Leading Drivers in Daytona Race," *New York Times,* January 30, 1966.
Mario: Brock Yates, "Mario on the Move," *Car and Driver,* August 1966.
Just give me: Mario Andretti, in discussion with the author.
We don't even know: "Ford, Ford, Ford," *Newsweek,* February 21, 1966.
193 *We had confidence in:* "Miles, Ruby Drive Ford to Victory," *Los Angeles Times,* February 7, 1966.
One of the most perfect: "Fords 1-2-3 in Daytona," *Chicago Tribune,* February 7, 1966.
I am proud for: "Double the Fun," *Time,* February 18, 1966.
Tears drip down: John Surtees, in discussion with the author.
Enzo Ferrari's rivalry with: John Surtees, *John Surtees: World Champion* (Surrey, U.K.: Hazleton Publishing, 1991), 183.
194 *They winched him up off:* Surtees, interview.
Would you consider: Surtees, *John Surtees,* 182.
You have to get straight: Bernard Cahier and Paul-Henri Cahier, *Grand Prix Racers: Portraits of Speed* (Minneapolis, MN: Motorbooks, 2008), 13.
195 *On March 15, he shattered:* "Surtees Sets Track Mark," *New York Times,* March 17, 1966.
On March 17, he went: "Surtees Smashes Third Monza Mark," *Washington*

Post, March 18, 1966. (As confirmed by lap times and in an interview with Surtees, the headline here contains a factual error. Monza should be Modena.)

196 *Lardi was put in charge:* Piero Ferrari, taped interview (translated by Ferrari liaison Matteo Sardi).

197 *Ferrari made frequent trips:* This story is covered thoroughly in Brock Yates, *Enzo Ferrari: The Man, the Cars, the Races, the Machine* (New York: Doubleday, 1991), 242.

Sunday there's Sebring: Franco Gozzi, *Memoirs of Enzo Ferrari's Lieutenant* (Milan: Giorgio Nada Editore, 2002), 108.

19. Blood on the Track

199 *For over a half century:* Ralph Nader, *Unsafe at Any Speed: The Designed-In Dangers of the American Automobile* (New York: Grossman Publishers, 1972), lxxxviv.

200 *President Johnson called:* "Auto Safety Measure's Provisions Disclosed," *Washington Post,* February 14, 1966.

Henry II hadn't read: Booton Herndon, *Ford: The Unconventional Story of a Powerful Family, a Mighty Industry and the Extraordinary Men Behind It All* (New York: Avon Books, 1970), 216.

If you take this: Herndon, *Ford,* 222.

201 *Waving a hammer at Ken Miles:* Leo Levine, *Ford: The Dust and the Glory,* Volume 1 (Warrendale, PA: Society of Automotive Engineers, 2000), 543.

Andretti made it back: Mario Andretti, in discussion with the author.

You've got it won: Carroll Shelby, in discussion with the author.

202 *Lloyd! You won!:* Lloyd Ruby, in discussion with the author. Also, Dave Friedman, *Shelby GT40* (St. Paul, MN: Motorbooks, 2006), 91.

Yes, yes, yes: Franco Gozzi, *Memoirs of Enzo Ferrari's Lieutenant* (Milan: Giorgio Nada Editore, 2002), 109.

An oscillograph spit out: Ron Wakefield, "Technical Analysis: Portrait of the Le Mans Winner," *Cobra, Shelby & Ford GT40 1962–1992* (Surrey, U.K.: Brooklands Books, n.d.), 44–45.

203 *John, you've got to:* John Surtees, in discussion with the author.

It read 3:46.8: Ford Motor Company Intra-Company Communication, May 5, 1966, minutes to Le Mans Committee meeting.

I hope he can sort: Dave Friedman, *Shelby GT40* (St. Paul, MN: Motorbooks, 2006), 96.

The longest hours: Walter Hayes, *Henry: A Life of Henry Ford II* (New York: Grove Weidenfeld, 1990), 95.

204 *Did you know him:* Hayes, *Henry,* 95.

Walt's subsequent laps: Ford Motor Company Intra-Company Communication, May 5, 1966, minutes to Le Mans Committee meeting.

20. The Blowout Nears

205 *You are the most:* Ken W. Purdy, "Stirling Moss: A Nodding Acquaintance with Death," *Playboy,* September 1962.

I'd like you to: John Wyer, *That Certain Sound* (Somerset, U.K.: Foulis Motoring Book, 1981), 154.

206 *John's idea of the perfect:* Denis Jenkinson, "Il Grande John," *Ferrari: The Man, the Machines,* ed. by Stan Grayson (Princeton, NJ: Automobile Quarterly Publications, 1975), 279.

207 *"[Ferrari] has no other satisfaction":* Griffith Borgeson, "The Great Agitator," *Ferrari: The Man, the Machines,* ed. Stan Grayson (Princeton, NJ: Automobile Quarterly Publications, 1975), 41.

"He never, in sickness": Walter Hayes, *Henry: A Life of Henry Ford II* (New York: Grove Weidenfeld, 1990), xiv.

Considering canceling the race: "Safety Campaign Urged On Le Mans Organizers," *London Times,* April 27, 1966.

Ferrari had received a note: Franco Gozzi, in discussion with the author.

We know that nothing: "Ferrari Says European Racing Doomed Before U.S. Steamroller," *Washington Post,* March 24, 1966.

208 *Incompetent dictator!:* Gozzi, interview. Also: Franco Gozzi, *Memoirs of Enzo Ferrari's Lieutenant* (Milan: Giorgio Nada Editore, 2002), 80.

Hit me like whiplashes: Gozzi, *Memoirs,* 80.

We make 12-cylinder: John Surtees, in discussion with the author. Also: John Surtees, *John Surtees: World Champion* (Surrey, U.K.: Hazleton Publishing, 1991), 183.

209 *In Surtees's paranoid state:* Gozzi, *Memoirs,* 80.

We will decide: Gozzi, interview.

Contact Mario Andretti: Gozzi, *Memoirs,* 81.

210 *Are you asking me:* "Critic of Auto Industry's Safety Standards Says He Was Trailed and Harassed; Charges Called Absurd," *New York Times,* March 6, 1966.

A thorough survey on: "Ferrari Says Speeding Does Not Cause Mishap," *New York Times,* May 10, 1966.

You will agree: "Henry Ford II Scores Auto-Safety Critics, Warns Congress Against 'Irrational' Laws," *Wall Street Journal,* April 18, 1966.

211 *Talk of nervous breakdowns:* Dave Friedman, in discussion with the author.

212 *The schematic utilized:* Robert Hogle, "Mark II-GT Ignition and Electrical System," Society of Automotive Engineers paper #670068.

Boeing 707 aircraft: Hogle, "Mark II-GT Ignition and Electrical System."

3.6 hours in the act: H.L. Gregorich and C.D. Jones, "Mark II—GT Transaxles," Society of Automotive Engineers paper #670069.

12,597,900 ft/lbs: Joseph J. Ihnacik Jr. and Jerome F. Meek, "Mark II GT Sports Car Disc Brake System," Society of Automotive Engineers paper #670070.

Indoor Laboratory Le Mans: B.F. Brender, C.J. Canever, I.J. Monti, and J.R. Johnson, "Laboratory Simulation, Mark II—GT Powertrain," Society of Automotive Engineers paper #670071.

213 *Behave like a human driver:* Brender et al., "Laboratory Simulation, Mark II—GT Powertrain."

That's up to you: "Ken Miles: The Victor Belongs to the Spoils," *Los Angeles Times,* August 22, 1966.

214 *Miles the Man for Le Mans:* "Miles the Man for Le Mans After Win in Daytona 24-Hour Enduro," *Los Angeles Times,* February 10, 1966.
Pushed American prestige: "Miles and Ruby Push American Prestige to Top with Victory in Tragedy-Studded Sebring Race," *Hartford Courant,* March 28, 1966.
I am a mechanic: Art Evans, *Ken Miles* (Redondo Beach, CA: Photo Data Research, 2004), 79.
I feel our chances: "Miles the Man."

215 *He had a dream of returning:* Mario Andretti, in discussion with the author.
Please try to remember: Andretti, interview.
At the moment: Gozzi, *Memoirs,* 81.
As well as going: Gozzi, *Memoirs,* 81.

216 *This is the most stupid:* Surtees, interview.
You go and make: Gozzi, *Memoirs,* 83.

217 *When you tell me how:* Surtees, *John Surtees,* 186.
He won: Gozzi, *Memoirs,* 83.

21. The Flag Drops

218 *This racetrack is:* "Many Top Drivers Afraid of Le Mans—and There are Many Good Reasons," *Detroit News,* June 17, 1966.
I knew I had to: "Foyt Tells How He Escaped Car," *Los Angeles Times,* June 6, 1966.

219 *Phil Hill had a falling out:* Phil Hill, filmed interview, *Carroll Shelby: The Man and His Car* (Duke DVD, 2004).
Money was too good: "Newcomer at Dangerous Le Mans," *New York Times,* June 16, 1966.

220 *We have our own medical:* "Many Top Drivers."

221 *The road death:* "Road Safety," *London Sunday Telegraph,* 1966 (clipping from Ford archives, ACC 1790, box 16).
Did you hear: "Newcomer at Dangerous Le Mans."
We're all here to make money: "Newcomer at Dangerous Le Mans."

222 *To do something well:* Eoin Young, *McLaren Memories* (Newbury Park, CA: Haynes Publishing, 2005), 157.

223 *Pussyfooted:* Eoin Young, *Forza Amon* (Newbury Park, CA: Haynes Publishing, 2003), 84.
We're not screwing: Young, *Forza Amon,* 84.
The only way we're: John Surtees, in discussion with the author.

224 *Making passes at her:* John Surtees, *John Surtees: World Champion* (Surrey, U.K.: Hazleton Publishing, 1991), 186.
Why is Scarfiotti's: Surtees, interview; Franco Gozzi, in discussion with the author. Also: Surtees, *John Surtees,* 187.

225 *I'm the fastest man:* Surtees, interview. Also: Surtees, *John Surtees,* 187.
The driver sent a telex: Gozzi, interview.
If Dragoni has decided: "Surtees Quits Le Mans Race After a Row," *London Daily Express,* June 17, 1966.

226 *Dragoni and I have:* "Surtees Quits Le Mans Race After a Row"; "Surtees Quits

Le Mans in Dispute," *Detroit News*, June 17, 1966; ABC *Wild World of Sports* footage, June 18, 1966; Surtees, interview.

227 *Well, Roy:* Roy Lunn, in discussion with the author.

228 *Let's drop the subject":* "Le Polemiche Si Attenuano Parte la 24 Ore di Le Mans," *Corriere della Sera*, June 18, 1966.

229 *Ford's sixteen drivers reported:* Young, *McLaren Memories*, 205.
Miles would race: Dave Friedman, *Remembering the Shelby Years—1962–1969* (Los Angeles: Carroll Shelby Children's Foundation, 1998), 240.
I appreciate that: Young, *McLaren Memories*, 205.
He then assigned lap speeds: Shelby American Inc.—Engineering & Development Report, June 21, 1966.

230 *For maximum durability:* Shelby American Inc.—Engineering & Development Report, June 21, 1966.
Buckle their belts: Mario Andretti, in discussion with the author.
Let's get it done: Young, *McLaren Memories*, 205.
I am Italian!: "The Year of the Ford," *Newsweek*, July 4, 1966.
Ford is an international: "Here to Show the World," *Sports Illustrated*, June 27, 1966.
Staring down at this Italian: ABC *Wide World of Sports* footage, June 18, 1966.
You better win: Dave Friedman, *Shelby GT40* (St. Paul, MN: Motorbooks, 1995), 112.

231 *It remained there:* Dave Friedman, in discussion with the author.

22. Le Mans – Record Pace

232 *Deuce was right there:* Edsel Ford II, in discussion with the author.

233 *The #1 Ford in at 5:26 P.M.:* The pit stop facts in this paragraph are all from Shelby American Inc.—Engineering & Development Report, June 21, 1966.
It's up to you: Eoin Young, *Forza Amon* (Newbury Park, CA: Haynes Publishing, 2003), 85.
We are witnessing: ABC *Wide World of Sports* footage, June 18, 1966.

234 *At 6:47 P.M.:* Shelby American Inc.—Engineering & Development Report, June 21, 1966.
They were so fast: Eoin Young, *McLaren Memories* (Newbury Park, CA: Haynes Publishing, 2005), 210.
Is it okay?: Young, *McLaren Memories*, 206.

235 *There you can see:* ABC *Wide World of Sports* footage, June 18, 1966.

236 *Like expectant fathers:* Brock Yates, "Le Mans 24 Hours," *Car and Driver*, September 1966.
The old man is: Yates, "Le Mans 24 Hours."

23. The Most Controversial Finish in Le Mans History

238 *At 4:10 A.M.:* Shelby American Inc.—Engineering & Development Report, June 21, 1966.

There was some talk: Dave Friedman, *Shelby GT40* (St. Paul, MN: Motorbooks, 1995), 105.

239 *Ken's been leading:* Friedman, *Shelby GT40,* 115.

240 *The officials say:* Leo Levine, *The Dust and the Glory,* Volume 1 (Warrendale, PA: Society of Automotive Engineers, 2000), 557.

Beebe had a chat: Carroll Shelby, in discussion with the author.

Why don't you bring: Eoin Young, *McLaren Memories* (Newbury Park, CA: Haynes Publishing, 2005), 207.

Who's supposed to win?: Friedman, *Shelby GT40,* 113.

241 *So ends my contribution:* Friedman, *Shelby GT40,* 112.

I don't know what: Friedman, *Shelby GT40,* 112.

Leo, the officials now: Levine, *The Dust and the Glory,* 558.

242 *The car was cruising:* Brock Yates, "Le Mans 24 Hours," *Car and Driver,* September 1966.

243 *I think I've been:* Friedman, *Shelby GT40,* 114.

Let's face it: Yates, "Le Mans 24 Hours."

244 *From there on:* Young, *McLaren Memories,* 206.

He just couldn't get over it: Albert Bochroch, *Americans at Le Mans* (Tucson, AZ: Aztex Corp., 1976), 147.

I considered we had won: "Ken Miles Recounts 'Defeat' at Le Mans," *Los Angeles Times,* July 3, 1966.

245 *Please be careful how:* "Le Mans Race Tribute to Miles," *Los Angeles Times,* June 8, 1967.

We don't want to buy: "A Le Mans non ha perso la tecnica Ferrari ma la sua 'dimensione,'" *Corriere della Sera,* June 21, 1966.

24. The End of the Road

246 *I felt I ought not:* John Surtees, *John Surtees: World Champion* (Surrey, U.K.: Hazleton Publishing, 1991), 187.

People rose to their feet: Eoin Young, in discussion with the author.

247 *He made comments that:* Surtees, *John Surtees,* 187.

A "divorce": John Surtees, in discussion with the author.

248 *Oh my God:* Leo Levine, *Ford: The Dust and the Glory,* Volume 1 (Warrendale, PA: Society of Automotive Engineers, 2000), 596.

I remember seeing the car: Art Evans, *Ken Miles* (Redondo Beach, CA: Photo Data Research, 2004), 111.

Where is he?: Evans, *Ken Miles,* 111.

249 *We don't have to feel:* "Ken Miles: The Victor Belongs to the Spoils," *Los Angeles Times,* August 22, 1966.

We really don't know: "Ford Will Probe Death-Car Crash," *Los Angeles Times,* August 19, 1966.

There will never be another: "Le Mans Race Tribute to Miles," *Los Angeles Times,* June 8, 1967.

The evidence is that: "Engineers Probe Causes of Miles' Fatal Crackup," *Detroit News,* August 27, 1966.

250 *We are here today . . . Life brilliant and vital:* Evans, *Ken Miles,* 108–109.

Epilogue

251 *Dear friends:* Official Ferrari 1966 Yearbook.

252 *I drenched Henry Ford:* Dan Gurney, in discussion with the author.
All-American Victory: "Foyt and Gurney Drive Ford Mark IV to Victory in 24 Hours of Le Mans," *New York Times,* June 12, 1967.

253 *He enjoyed beautiful:* "He Enjoyed Beautiful Women, Strong Drink," *Detroit News,* September 30, 1987.
His insistence on: "Dynastic Heir Who Rescued Ford Firm from Bankruptcy," *Financial Times,* September 30, 1987.
Fans carried me high!: John Surtees, in discussion with the author.

254 *I don't think Ferrari:* Surtees, interview.
We would not be: ABC's *Wide World of Sports,* Sports Highlights of the 1960s (December 27, 1969), in the collection of the Paley Center for Media in New York.

254 *I saw my friends:* "Deaths that Drove Stewart to Transform F1 Legend Ranks Safety Reforms as His Greatest Success," *Scotland on Sunday,* October 28, 2007.

255 *You wouldn't believe:* Carroll Shelby, in discussion with the author.
I'll forever be: Shelby, interview.

256 *It broke my heart:* Shelby, interview.

INDEX